In Vivo Models to Study Angiogenesis

In Vivo Models to Study Angiogenesis

Domenico Ribatti
University of Bari Medical School, Bari, Italy
National Cancer Institute "Giovanni Paolo II", Bari, Italy

ACADEMIC PRESS

An imprint of Elsevier

Academic Press is an imprint of Elsevier
125 London Wall, London EC2Y 5AS, United Kingdom
525 B Street, Suite 1800, San Diego, CA 92101-4495, United States
50 Hampshire Street, 5th Floor, Cambridge, MA 02139, United States
The Boulevard, Langford Lane, Kidlington, Oxford OX5 1GB, United Kingdom

British Library Cataloguing-in-Publication Data
A catalogue record for this book is available from the British Library

Library of Congress Cataloging-in-Publication Data
A catalog record for this book is available from the Library of Congress

ISBN: 978-0-12-814020-8

Working together
to grow libraries in
developing countries

www.elsevier.com • www.bookaid.org

Publisher: Jonathan Simpson
Acquisition Editor: Glyn Jones
Editorial Project Manager: Charlotte Rowley
Production Project Manager: Mohanapriyan Rajendran
Cover Designer: MPS

Typeset by MPS Limited, Chennai, India

CONTENTS

The cardiovascular system plays a crucial role in vertebrate development and homeostasis. During embryonic development, vasculature is formed by both vasculogenesis and angiogenesis. Vasculogenesis, which consists of *de novo* vessel formation from angioblasts, provides the nascent vascular network, particularly during early embryonic life. Angiogenesis, which refers to expansion of a preexisting vascular bed through sprouting, bridging, and intussusceptive growth, intervenes mostly during later stages of embryogenesis (Ribatti, 2006). Bone marrow−derived stem and endothelial progenitor cells can in principle contribute to tissue repair by the induction of neo-angiogenesis (Fig. 1).

Several genetic and epigenetic mechanisms are involved in the early development of the vascular system and there is an extensive literature on the genetic background and the molecular mechanisms responsible for blood vessel formation. Evidence is emerging that blood vessels themselves have the ability to provide instructive regulatory signals to surrounding nonvascular target cells during organ development. Thus endothelial cell signaling is believed to promote fundamental cues for cell fate specification, embryo pattering, organ differentiation, and postnatal tissue remodeling. The ability of vessels to influence surrounding nonvascular cells may depend on the intrinsic heterogeneity of endothelial cells. Indeed, one of the most interesting theoretical perspectives and practical applications of endothelial cell signaling is the possibility for these cells to maintain their inductive potential during adult life.

Starting with the hypothesis of Judah Folkman (1933−2008) (Fig. 2) that tumor growth is angiogenesis dependent, this area of research has a solid scientific foundation. Folkman found a revolutionary new way to think about cancer. He postulated that in order to survive and grow, tumors require blood vessels and that by cutting off the blood supply, a cancer could be starved into remission (Ribatti, 2008).

Solid tumor growth occurs by means of an avascular phase followed by a vascular phase (Ribatti et al., 1999). Assuming that such

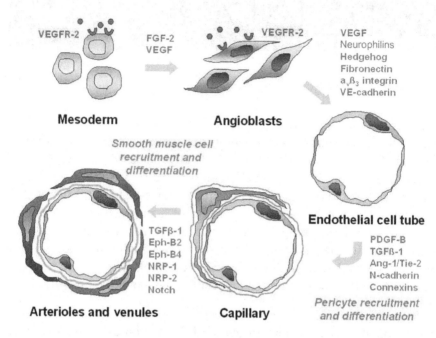

Figure 1 *The earliest blood vessels in the embryo originate from mesodermal cells that are specified into angioblasts most likely in response to FGF2 and VEGF signals. Angioblasts begin to differentiate into endothelial cells and assemble into tubes, principally as a result of VEGF signals from surrounding tissues and the expression of intercellular and cell-matrix adhesion molecules. Endothelial cell tubes are soon stabilized by pericytes recruited from the surrounding mesenchyme to form early capillaries. In microvessels, PDGF and TGFβ-1 signals are involved in the recruitment of pericytes. In larger vessels, arterioles and venules, the vascular wall is made up of endothelial cells and smooth muscle cells, which are recruited mainly through the Tie-2 and Ang-1 receptor—ligand pair, although Neuropilins and Notch pathway are also involved in mural cell formation. Ephrin-B2 and Ephrin-B4 are implicated in arterial and venous endothelial cell specialization, respectively.* Reproduced from Ribatti, D., Vacca, A., Dammacco, F., 1999. The role of the vascular phase in solid tumor growth: a historical review. Neoplasia 1, 293—302 (Ribatti et al., 2009).

Figure 2 *A port trait of Judah Folkman.*

growth is dependent on angiogenesis and that this depends on the release of angiogenic factors, the acquisition of an angiogenic ability can be seen as an expression of progression from neoplastic transformation to tumor growth and metastasis. Practically all solid tumors, including those of the colon, lung, breast, cervix, bladder, prostate, and pancreas, progress through these two phases. The role of angiogenesis in the growth and survival of leukemias and other hematological malignancies has only become evident since 1994 thanks to a series of studies demonstrating that progression in several forms is clearly related to their degree of angiogenesis (Vacca and Ribatti, 2006).

Tumor angiogenesis is linked to a switch in the balance between positive and negative regulators, and mainly depends on the release by neoplastic cells of specific growth factors for endothelial cells that stimulate the growth of the host's blood vessels or the downregulation of natural angiogenesis inhibitors (Fig. 3) (Ribatti et al., 2007a). In normal tissues, vascular quiescence is maintained by the dominant influence of endogenous angiogenesis inhibitors over angiogenic stimuli.

The blood vessels of tumors display many structural and functional abnormalities (Ribatti et al., 2007b). Their unusual leakiness, potential for rapid growth and remodeling, and expression of distinctive surface

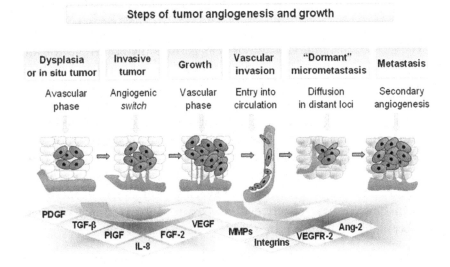

Figure 3 Steps of tumor angiogenesis and growth. Reproduced from Ribatti, D., Vacca, A., 2008. Overview of tumor angiogenesis. In: Angiogenesis. An Integrative Approach From Science to Medicine. Springer, New York, pp. 161–168.

molecules mediate the dissemination of tumor cells in the bloodstream and maintain the tumor microenvironment. Like normal blood vessels, they consist of endothelial cells, pericytes, and their enveloping basement membrane. Common features, regardless of their origin, size, and growth pattern, include the absence of a hierarchy, the formation of large-caliber sinusoidal vessels, and a markedly heterogeneous density.

There is increasing evidence to support the view that angiogenesis and inflammation are mutually dependent (Ribatti and Crivellato, 2009) During inflammatory reactions, immune cells synthesize and secrete pro-angiogenic factors that promote neovascularization. On the other hand, the newly formed vascular supply contributes to the perpetuation of inflammation by promoting the migration of inflammatory cells to the site of inflammation (Ribatti and Crivellato, 2009).

Tumor cells are able to secrete pro-angiogenic factors as well as mediators for inflammatory cells. They produce indeed angiogenic cytokines, which are exported from tumor cells or mobilized from the extracellular matrix. As a consequence, tumor cells are surrounded by an infiltrate of inflammatory cells. These cells communicate via a complex network of intercellular signaling pathways, mediated by surface adhesion molecules, cytokines and their receptors. Immune cells cooperate and synergize with stromal cells as well as malignant cells in stimulating endothelial cell proliferation and blood vessel formation (Fig. 4).

The genetic instability of tumor cells permits the occurrence of multiple genetic alterations that facilitate tumor progression and metastasis, and cell clones with diverse biological aggressiveness may coexist within the same tumor. These two properties allow tumors to acquire resistance to cytotoxic agents. Whereas conventional chemotherapy, radiotherapy, and immunotherapy are directed against tumor cells, antiangiogenic therapy is aimed at the vasculature of a tumor and will either cause total tumor regression or keep tumors in a state of dormancy. Even though numerous compounds inhibit angiogenesis, few of them have proved effective in vivo, and only a couple of agents have been able to induce tumor regression.

Beginning in the 1980s, the biopharmaceutical industry began exploiting the field of antiangiogenesis for creating new therapeutic compounds for modulating new blood vessel growth in

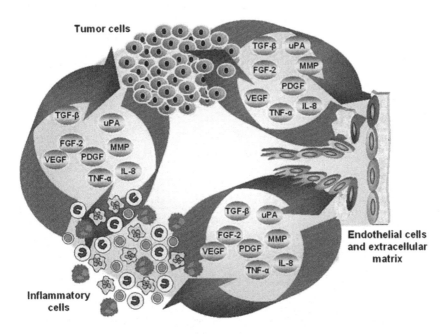

Figure 4 Interplay between tumor cells, inflammatory cells, and extracellular matrix in inducing angiogenic response. Reproduced from Ribatti, D., Vacca, A., 2008. Overview of tumor angiogenesis. In: Angiogenesis. An Integrative Approach From Science to Medicine. Springer, New York, pp. 161–168.

angiogenesis-dependent diseases, and the number of patients receiving antiangiogenic therapies for cancer treatment has progressively increased.

Antiangiogenic agents may be classified as synthetic angiogenesis inhibitors and endogenous angiogenesis inhibitors (Ribatti, 2009). Some inhibit endothelial cells directly (blocking endothelial cells from proliferating, migrating, or increasing their survival in response to pro-angiogenic molecules), while others inhibit the angiogenesis signaling cascade (blocking the activity of one, two, or a broad spectrum of pro-angiogenic proteins and/or their receptors) or block the ability of endo-thelial cells to break down the extracellular matrix, which is required to allow endothelial cells to migrate into the surrounding tissues and proliferate into new vessels. Angiogenesis inhibitors may also be char-acterized by the degree of blocking potential: drugs that block one main angiogenic protein, drugs that block two or three main angio-genic proteins, or drugs that block a range of angiogenic regulators.

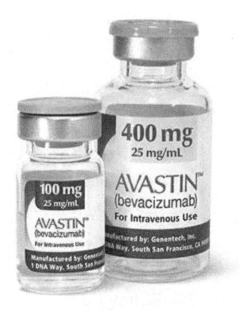

Figure 5 Bevacizumab (Avastin) is administered as an intravenous injection.

The most promising antiangiogenic agents that are in clinical development at this moment include bevacizumab (Avastin) (Fig. 5), the humanized antimonoclonal antibody anti-vascular endothelial growth factor (VEGF) approved for use in combination with cytotoxic agents, as well as small molecules receptor tyrosine kinase inhibitors (RTKIs), approved as single agents, and including sunitinib, an oral inhibitor of VEGF receptor-2 (VEGFR-2), platelet derived growth factor receptor (PDGFR), FlLT-3, and c-KIT, and sorafenib, an inhibitor of the Faf/MEK/Erk and the VEGFR and PDGFR signaling pathways.

Bevacizumab is considered to be the first specific angiogenesis inhibitor for clinical oncology. However, the results from clinical trials have not shown the dramatic antitumor effects that were expected following preclinical studies, which revealed a much higher efficacy of these types of agents in animal models. Patients with different types of tumors respond differently to antiangiogenic therapy. Additionally, preclinical and clinical data have shown the possibility that tumors may acquire resistance to antiangiogenic drugs or may escape antiangiogenic therapy via compensatory mechanisms (Ribatti, 2016). Multiple angiogenic molecules may be produced by tumors, and tumors at different stages of development may depend on different angiogenic factors for their blood supply. Therefore, blocking a single angiogenic molecule might have little or no impact on tumor growth.

Angiogenesis is a typical example of a scientific field of the biomedical sciences which necessitates both in vivo and in vitro investigation. As pointed out by Auerbach et al. (1991) "perhaps the most consistent limitation (to progress in angiogenesis research) has been the availability of simple, reliable, reproducible, quantitative assays of the angiogenesis response." Continuous monitoring of angiogenesis in vivo is required for the development and evaluation of drugs acting as stimulators or suppressors of angiogenesis.

In vivo models of angiogenesis carry out all the steps of angiogenesis and vessel maturation to produce fully functional vascular networks. The classical assays for studying angiogenesis in vivo include the hamster cheek pouch, rabbit ear chamber, dorsal skin and air sac, the chick embryo chorioallantoic membrane (CAM), and iris and avascular corneal of rodent eye. Several new models have been introduced including subcutaneous implantation of various three-dimensional substrates including polyester sponge, polyvinyl-alcohol foam disc covered on both sides with a Millipore filter (the disc angiogenesis system), and Matrigel, a basement membrane–rich extracellular matrix. The most reliable in vivo angiogenesis techniques use the CAM and the rabbit cornea. In contrast to the CAM assay, the other in vivo techniques are more complex, consume a large quantity of angiogenic factors, are not feasible for numerous samples, and are expansive.

A "gold standard" angiogenesis assay has yet to be developed. A single in vivo model is inadequate to fully investigate the process of angiogenesis as there are variations between species, organ sites, and specific microenvironment. Therefore a combination of assays is required.

The timing, concentration, location, and mode of treatment administration are critical in all animal experiments regardless of species. Moreover, the concentrations of cytokines required to observe angiogenesis in any in vivo assay will vary from the concentrations of angiogenic growth factors used to stimulate endothelial cells in in vitro assays.

There has been an exponential increase in the sophistication of in vivo imaging techniques including the availability of Magnetic

Resonance Imaging (MRI), Computed Tomography (CT), and Positron Computed Tomography (PCT) facilities for scanning small animals, and the advent of confocal and multi-photon microscopy enabling fine structure imaging in situ.

The in vivo assays of angiogenesis have enabled to make up important progress in elucidating the mechanism of action of several angiogenic factors and inhibitors. It is reasonable to reserve the term "angiogenic factor" for a substance which produces new capillary growth in an in vivo assay. A variety of animal models have been described to provide more quantitative analysis of in vivo angiogenesis and to characterize pro- and antiangiogenic molecules. The principal qualities of the in vivo assays are their low cost, simplicity, reproducibility, and reliability which, in turn, among the different in vivo assays are important determinants dictating the choice of method.

However, they are also very sensitive to environmental factors, not readily accessible to biochemical analysis and their interpretation is frequently complicated by the fact that the experimental condition inadvertently favors inflammation, and that under these conditions the angiogenic response is elicited indirectly, at least in part, through the activation of inflammatory or other nonendothelial cells. On the basis of these limitations, ideally two different assays should be performed in parallel to confirm the angiogenic or antiangiogenic activities of test substances. The predictive value of angiogenesis assays remains to be established and results must be interpreted with care.

REFERENCES

Auerbach, R., Auerbach, W., Polakowski, I., 1991. Assays for angiogenesis: a review. Pharmacol Ther 51, 1–11.

Ribatti, D., 2006. Genetic and epigenetic mechanisms in the early development of the vascular system. J Anat 208, 139–152.

Ribatti, D., 2008. Judah Folkman, a pioneer in the study of angiogenesis. Angiogenesis 11, 3–10.

Ribatti, D., 2009. Endogenous inhibitors of angiogenesis. an historical review. Leukemia Res 33, 638–644.

Ribatti, D., Crivellato, E., 2009. Immune cells and angiogenesis. J Cell Mol Med 13, 2822–2833.

Ribatti, D., 2016. Tumor refractoriness to anti-VEGF therapy. Oncotarget 19, 46668–46677.

Ribatti, D., Vacca, A., 2008. Overview of tumor angiogenesis. In: Figg, W.D., Folkman, J. (Eds.), Angiogenesis. An Integrative Approach From Science to Medicine. Springer, New York, pp. 161–168.

Ribatti, D., Vacca, A., Dammacco, F., 1999. The role of the vascular phase in solid tumor growth: a historical review. Neoplasia 1, 293–302.

Ribatti, D., Nico, B., Crivellato, E., et al., 2007a. The history of the angiogenic switch comcept. Leukemia 21, 44–52.

Ribatti, D., Nico, B., Crivellato, E., et al., 2007b. The structure of the vascular networks of tumors. Cancer Lett 248, 18–23.

Ribatti, D., Nico, B., Crivellato, E., 2009. Morphological and molecular aspects of physiological vascular morphogenesis. Angiogenesis 12, 101–111.

Vacca, A., Ribatti, D., 2006. Bone marrow angiogenesis in multiple myeloma. Leukemia 20, 193–199.

CHAPTER 1

The Chick Embryo Chorioallantoic Membrane

The chick embryo chorioallantoic membrane (CAM) assay facilitates the testing of multiple samples and the generation of dose–dilution curves, and has been used to identify almost all of the known angiogenic factors. The allantois is an extraembryonic membrane, derived from the mesoderm in which primitive blood vessels begin to take shape on day 3 of incubation. On day 4, the allantois fuses with the chorion and forms the chorioallantois (Figs. 1.1 and 1.2). Until day 8 of incubation, primitive vessels continue to proliferate and to differentiate into an arteriovenous system and thus originate a network of capillaries that migrate to occupy an area beneath the chorion and mediate gas exchanges with the outer environment. The CAM vessels grow rapidly up to day 11, after which the endothelial cell mitotic index decreases just as rapidly, and the vascular system attains its final setup on day 18 of incubation, just before hatching (Ausprunk et al., 1974).

Fuchs and Lindenbaum (1988) described six or seven generations of branches of the allantoic artery. The first five or six are located in a plane parallel to the CAM surface and deep into the vein, which has a similar distribution. The fifth and sixth generations change direction, passing almost vertically in the two-dimensional (2D) capillary plexus (Fig. 1.3). Three types of order exhibited by the CAM vessels are considered: (1) positional order, associated with the arrangement of the network in the membrane; (2) topological order, associated with the homogeneity of branching occurring along the structure; (3) orientation order, associated with the orientation of the branches forming the network. They are quantitatively evaluated starting from the image of the binary skeleton (Fig. 1.4) (Guidolin et al., 2004).

De Fouw et al. (1989) have shown rapid extension of the CAM surface from 6 cm^2 at day 6 to 65 cm^2 at day 14. The number of feed vessels increased (2.5- and 5-fold for precapillary and postcapillary vessels), predominantly due to growth and remodeling after day 10.

In Vivo Models to Study Angiogenesis. DOI: http://dx.doi.org/10.1016/B978-0-12-814020-8.00001-9

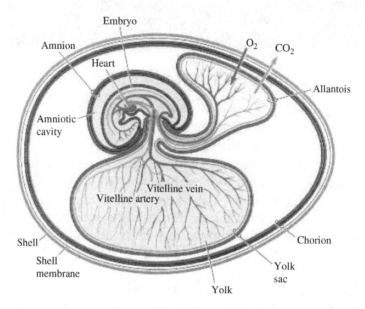

Figure 1.1 Schematic drawing of the embryo, amnion, yolk sac, allantois, and chorion with the blood vessels in the yolk sac, the allantois, and the embryo.

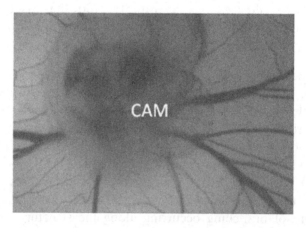

Figure 1.2 Macroscopic in vivo features of the CAM at day 5 of incubation. Reproduced from Ribatti, D., 2014. The chick embryo chorioallantoic membrane as a model for tumor biology. Exp. Cell. Res. 328, 314–324.

The growth of the vascular network in the CAM results in an increase of the O_2 diffusing capacity of the CAM in a sigmoid fashion, parallel to the change in O_2 uptake of the embryo. The O_2 consumption of the embryo increases slowly during the first week and a half of incubation (Rahn et al., 1979).

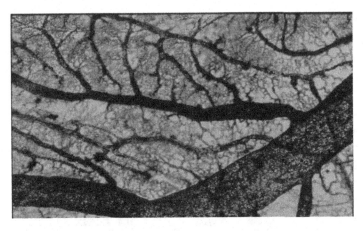

Figure 1.3 India ink injection of the vascular tree of the CAM. Six or seven generations of branches of the allantoic artery are described. The first five or six are located in a plane closely parallel to the CAM surface, and deep into the location of the veins which have a similar distribution. The fifth and sixth generation of blood vessels (i.e., the pre- or postcapillary venules) change direction, passing almost vertically into the 2D capillary plexus.

According to Schlatter et al. (1997), CAM vascularization undergoes three phases of development with both sprouting and intussusceptive microvascular growth (IMG). In the early phase, multiple capillary sprouts invade the mesenchyme, fuse, and form the primary capillary plexus. During the second phase, sprouts are no longer present and have been replaced by tissue pillars with expression of IMG. During the late phase, the growing pillars increase in size to form intercapillary meshes.

Immature blood vessels scattered in the mesoderm grow very rapidly until day 8 and give rise to a capillary plexus, which comes to be intimately associated with the overlying chorionic epithelial cells and mediates gas exchange with the outer environment. At day 14, the capillary plexus is located at the surface of the ectoderm adjacent to the shell membrane. Rapid capillary proliferation continues until day 11; thereafter, the endothelial cell mitotic index declines rapidly, and the vascular system attains its final arrangement on day 18, just before hatching (Ausprunk et al., 1974).

On day 4, all CAM vessels have the appearance of undifferentiated capillaries. Their walls consist of a single layer of endothelial cells lacking a basal lamina (Ausprunk et al., 1974). By day 8, the CAM displays small, thin-walled capillaries beneath the chorionic epithelium, and other vessels in the mesodermal layer, whose walls have a layer of

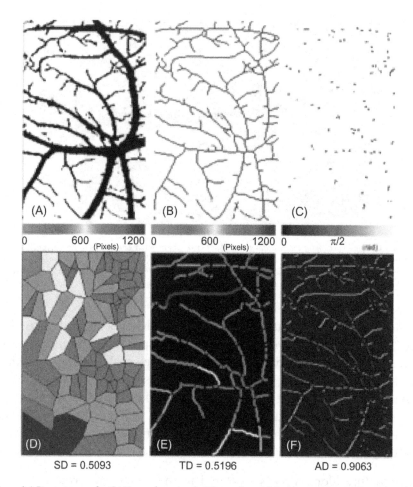

SD = 0.5093   TD = 0.5196   AD = 0.9063

Figure 1.4 Binary image of a CAM vascular network (A) together with its binary skeleton (B) and branching points (C). The corresponding Voronoi diagram is shown in (D), where cells were color coded according to their size. Skeleton segments connecting branching points are shown in (E) coded in colors according to their length. In (F), the pixels of the binary skeleton are shown in color coding for their orientation. At the bottom, values of the order parameters are reported. Reproduced from Guidolin, D., Nico, B., Mazzocchi, G., et al., 2004. Order and disorder in the vascular network. Leukemia 18, 1745–1750.

mesenchymal cells surrounding the endothelium and are completely wrapped by a basal lamina together with the endothelial cells (Fig. 1.5) (Ausprunk et al., 1974). On days 10–12, the capillaries are near the surface of the chorionic epithelium. The mesodermal vessels are now distinct arterioles and venules. The walls of arterioles contain one or two layers of mesenchymal cells and increased amounts of connective tissue surrounding them. Venules are surrounded by an incomplete investment of mesenchymal cells, which are to be developing

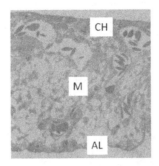

Figure 1.5 A semi-thin section of the CAM, showing the chorionic epithelium (CH), the intermediate mesenchyme (M), and the deep allantoic epithelium (AL). Reproduced from Ribatti, D., 2014. The chick embryo chorioallantoic membrane as a model for tumor biology. Exp. Cell Res. 328, 314—324.

smooth muscle cells and the walls of arterioles also develop a distinct adventitia containing fibroblast-like cells.

UTILIZATION OF THE CAM

Fertilized chick eggs are incubated at 37°C, at constant humidity. Any breed of chicken can be used. It is important that the eggs are free of the major pathogens that could affect embryo development and contaminate the laboratory environment and personnel. Chick embryos are rarely contaminated with bacteria but are easily contaminated with fungi. All procedures should be performed in a clean room with sterilized instruments and materials. On day 3 of incubation, aspirate 2—3 mL albumen at the more pointed end of the egg, so that the CAM can be detached from the shell itself, a window is cut into the shell with the aid of scissors, and the underlying CAM vessels are demonstrated (Fig. 1.6).

CAM was first used to study tumor angiogenesis by grafting tumor samples onto its surface. Different tumors and cell suspensions derived from tumors have been implanted on the CAM (Fig. 1.7) (Ribatti et al., 1990, 2010). Between 2 and 5 days after tumor cell inoculation, the tumor xenografts become visible, are supplied with vessels of CAM origin, and begin a phase of rapid growth. Knighton et al. (1977) investigated the time course of Walker 256 carcinoma specimens implanted on the CAM (Fig. 1.8). Without heparin, tumor angiogenesis is usually detectable in about 3 days. With only 2 units of heparin (12 μg), neovascularization appeared within 1 day (Taylor and

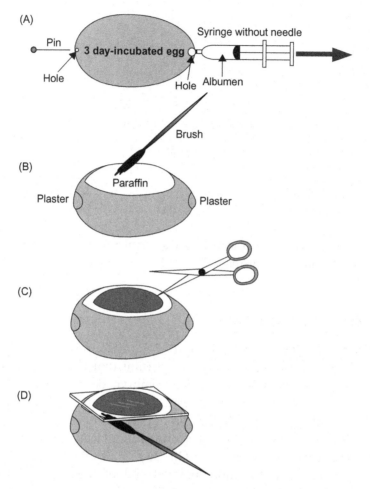

Figure 1.6 Preparation of the egg for the CAM assay. On day 3 of incubation, 2–3 mL of albumen are aspirated at the acute pole of the egg (A) to detach the developing CAM from the shell; the holes are closed with plaster. The upper surface of the egg is brushed on with paraffin (B) and cut with scissor kept parallel to the surface so as not to damage the embryo (C). The window is covered with a glass slide and sealed with paraffin (D).

Folkman, 1982). When heparin and cortisone were added together, angiogenesis was inhibited (Folkman et al., 1983).

Ausprunk et al. (1975) compared the behavior of tumor grafts to grafts of normal adult and embryo tissues. In tumor tissue, preexisting blood vessels in the tumor graft disintegrated within 24 h after implantation, and revascularization occurred by penetration of proliferating host vessels into the tumor tissue, and tumor vessels did not reattach to those of the host. In the embryo grafts, preexisting vessels did not

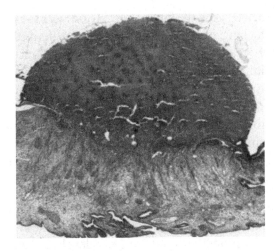

Figure 1.7 Highly vascularized CAM, 4 days after implantation of a bioptic specimen of a human B-cell non-Hodgkin's lymphoma. Reproduced from Ribatti, D., Vacca, A., Bertossi, M., et al., 1990, Angiogenesis induced by B-cell non-Hodgkin's lymphomas. Lack of correlation with tumor malignancy and immunologic phenotype. Anticancer Res. 10, 401–406.

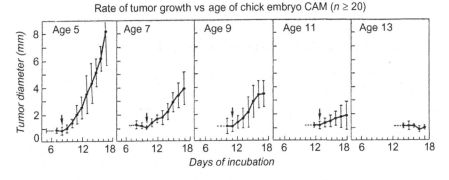

Figure 1.8 Implants of the Walker carcinoma on the CAM. Tumors did not exceed a mean diameter of 0.93 ± 0.29 mm during the prevascular phase (approximately 72 h). Rapid growth begins 24 h after vascularization and tumors reach a mean diameter of 8.0 ± 2.5 mm by 7 days. When tumor grafts from 1 to 4 mm were implanted on the 9-day CAM, grafts larger than 1 mm undergo necrosis and autolysis during the prevascular phase. Reproduced from Knighton, D.R., Ausprunk, D.H., Tapper, D., et al., 1977. Avascular and vascular phases of tumor growth in the chick embryo, Br. J. Cancer 35, 347–356.

disintegrate in the embryo grafts and anastomosed to the host vessels within 1–2 days, but with minimal or almost no neovascularization of the host vessels. In adult tissues, preexisting graft vessels disintegrated, did not stimulate capillary proliferation in the host, and there was no reattachment of their circulation with the host. Hagedorn et al. (2005) demonstrated that glioblastoma multiforme cells implanted on the CAM formed avascular tumors within 2 days, which progressed through vascular endothelial growth factor receptor-2 (VEGFR-2)-dependent angiogenesis. Blocking of VEGFR-2 and platelet derived growth factor receptor (PDGFR) signaling pathways with small-molecule receptor tyrosine kinase inhibitors blocked tumor growth. Moreover, gene regulation analysis during the angiogenic switch by oligonucleotide microarrays identified genes associated with tumor vascularization and growth. When human glioblastoma cells were inoculated on the CAM, tumors showed histology very similar to human glioblastoma, characterized by diffused pleiomorphic infiltrate of fibrillar stellate cells with neoangiogenesis, edema, and areas of necrosis (Durupt et al., 2012).

It should be noted that CAM was found to be ideal for investigating the tumor-induced angiogenesis, because at that point in time the host's immunocompetent system was not yet fully developed, hence the conditions for rejection were not yet established (Leene et al., 1973). The presence of T cells can be first detected at day 11 and of B cells at day 12 (Janse and Jeurissen, 1991), and by day 18 chicken embryos become immunocompetent (Jankovic et al., 1975; Weber and Mausner, 1977).

THE METASTASIS CHICK EMBRYO MODEL

This experimental model based on the grafting of human tumor cells on the CAM has provided valuable information regarding tumor cell intravasation of the chorionic epithelium, the mesenchyme below and the blood vessels, survival of tumor cells in the circulation, their arrest in the vasculature, extravasation, and proliferation in the distant organs. More than 80% of the injected cells survive in the microcirculation and have successfully extravasated by 1–3 days later, migrate through the mesenchyme and attach to arterioles, and migrate to the vicinity of pre-existing vessels (Koop et al., 1994). The changes in morphology of cancer cells arrested in the CAM microcirculation can be readily observed by in vivo microscopy and most of them survive without significant cell damage and complete extravasation (Chambers et al., 1992).

Comparative analysis of a series of tumor cell lines, unrelated in origin, showed that they intravasated to the vasculature with different efficiencies (Kim et al., 1998). Microscopic evaluation of tumor cells in the CAM revealed a dynamic cellular microenvironment in which tumor cells move through the tumor tissue as well as the adjacent stroma (Zijlstra et al., 2002).

Innovative studies of Ossowski (Kim et al., 1998; Ossowski and Reich, 1980; Yu et al., 1997) and Chambers (Chambers et al., 1992; Koop et al., 1996; MacDonald et al., 1992) have consolidated the CAM as a useful model to study metastasis. Several methods for semiquantitative analysis of metastasis in the chick embryo have been developed, including morphometric assessment of individual metastasized cells (Ossowski and Reich, 1980), detection of microscopic tumor colonies (MacDonald et al., 1992), detection of human urokinase plasminogen activator within secondary organs of the embryo (Ossowski and Reich, 1980), the use of green fluorescent protein, and in vivo videomicroscopy (Khokha et al., 1992; Koop et al., 1996). Because the human genome is uniquely enriched in Alu sequences, polymerase chain reaction (PCR)-mediated amplification of human-specific Alu sequences was used for semiquantitative detection of intravasated tumor cells in the CAM and within chicken tissues (Kim et al., 1998), followed by sensitive real-time Alu PCR assay (Mira et al., 2002; van der Horst et al., 2004; Zijlstra et al., 2002).

Ossowski introduced a new assay in which tumor cell suspensions, or fragments, were implanted on the CAM surface (Kim et al., 1998). The CAM surface was previously wounded and parts of the chorionic epithelium were damaged, exposing vascularized stroma. Tumor cells entered this tissue regardless of their invasive potential, but only cells capable of penetrating the blood vessel wall subsequently circulated and arrested in vessels of embryonic tissues.

Chambers (Chambers et al., 1992) compared the patterns of experimental metastasis in chick and mouse after intravenous injection of murine melanoma cells, and observed differences between the two models: (1) the number of tumors for a given number of cells injected is much higher in the chick than in the mouse; (2) B16-F1 tumors grew in most embryonic chick organs while their growth in the mouse was restricted primarily to the lungs; (3) B16-F1 and B16-F10 formed a comparable number of tumors in embryonic organs after intravenous injection in the chick, whereas B16-F10 formed more tumors in the lung than B16-F1 after intravenous injections into mice.

Lugassy and Barnhill (Lugassy and Barnhill, 2007; Lugassy et al., 2006) have demonstrated that, while melanoma tumor cells close to the tumor inoculation were completely cuffing some vessels, further from the tumor, melanoma cells were observed in small groups of cells along the outside of the vessel and at distance from the tumor, isolated fluorescent tumor cells, as well as small tumor masses, were observed along the vessels. They named this process as angiotropism of melanoma cells, i.e., the capability of some melanoma cells to migrate along the outside of vessels in a pericyte-like location.

The major advantages of the CAM model as an experimental animal model of metastasis are: (1) the chick embryo is naturally immunodeficient and can accept cancer cells regardless of their origin without immune response; (2) the changes in morphology of cancer cells arrested in the CAM microcirculation can be readily observed by in vivo microscopy (Chambers et al., 1992); (3) most cancer cells arrested in the CAM microcirculation survive without significant cell damage and complete extravasation (Chambers et al., 1992).

PROANGIOGENIC AND ANTIANGIOGENIC MOLECULES

CAM is also used to study different macromolecules developing angiogenic and antiangiogenic activity (Ribatti et al., 1996). To do so, inert synthetic polymers soaked with the macromolecule one wants to be tested are laid upon the CAM surface. Elvax 40 and Hydron are commonly used. They were first described and validated by Langer and Folkman (1976): both proved to be biologically inert when implanted onto the CAM and both were found to polymerize in the presence of the test substance, allowing its sustained release during the assay. When polymers are used in combination with an angiogenic substance, a vasoproliferative response will be recognizable 72–96 h after implantation: the response takes the form of increased vessel density around the implant, with the vessel radially converging toward the center-like spokes in a wheel (Ribatti et al., 1995). Conversely, when polymers combined with an antiangiogenic substance are tested the vessels become less dense around the implant after 72–96 h, and eventually disappear. Protamine sulfate was the first angiogenesis inhibitor to be effective when administered systemically (Taylor and Folkmam, 1982). Protamine prevented tumor-induced angiogenesis on the CAM and it also inhibited the growth of embryonic vessels: in its presence, avascular zones appeared in the 6- to 8-day-old CAM (Taylor and Folkman, 1982).

A fluid substance, or cell suspensions, can be directly inoculated into the cavity of the allantoic vesicle so that their activity covers the whole vascular area in a uniform manner (Ribatti et al., 1987; Gualandris et al., 1996).

The test substance can be dissolved in 0.5% methylcellulose which was then pipetted onto Teflon molds of 10 μL each and allowed to dry and to make disks of about 2 mm diameter which strongly adhere to the CAM surface (Taylor and Folkman, 1982). Culture coverslide glasses 4–5 mm in diameter on which the angiogenic factors were placed may be used (Wilting et al., 1991). Glasses were turned over and placed onto the CAM on day 9 of incubation and the angiogenic response was evaluated 96 h later. In another method, the testing substance is placed into a collagen gel between two parallel nylon meshes which align the capillaries for counting. The resulting "sandwich" is then placed upon the CAM on day 8 of incubation (Nguyen et al., 1994). A major advantage of this method is that it does not require histological sections, thus facilitating the screening of a large number of compounds.

Still other methods have been proposed whereby the CAM vascular networks can be displayed in greater detail, except that the embryo with the extraembryonic membranes and yolk sac must be transferred into an in vitro system during the early stages of development (day 3 or 4 of incubation). The system consists of a large diameter (10 mm) Petri dish (Fig. 1.9) (Auerbach et al., 1974), plastic pipe, plastic wrap

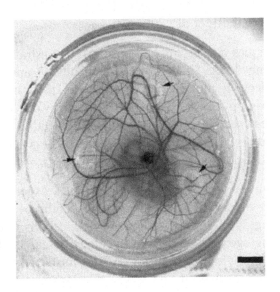

Figure 1.9 CAM assay in a Petri dish.

and tripod (Dunn et al., 1981), plastic cup and plastic wrap (Dugan et al., 1991), and plastic cups alone (Jakobson et al., 1989).

We have developed a new method for the quantitation of angiogenesis and antiangiogenesis in the CAM. Gelatin sponges treated with a stimulator or an inhibitor of blood vessel formation are implanted on growing CAM on day 8 (Ribatti et al., 1997, 2006). Blood vessels growing vertically into the sponge and at the boundary between sponge and surrounding mesenchyme are counted morphometrically on day 12. The newly formed blood vessels grow perpendicularly to the plane of the CAM inside the sponge, which does not contain preexisting vessels. The gelatin sponge is also suitable for the delivery of tumor cell suspensions, as well as of any other cell type, onto the CAM surface and the evaluation of their angiogenic potential (Ribatti et al., 1997, 1998; Vacca et al., 1998, 1999). As compared with the application of large amounts of a recombinant angiogenic cytokine in a single bolus, the use of cells implants that overexpress angiogenic cytokines allows the continuous delivery of growth factors, which is produced by a limited number of cells (as low as 10.000–20.000 cells per implant), thus mimicking more closely the initial stages of tumor angiogenesis and metastasis. Cells that overexpress fibroblast growth factor-2 (FGF-2) or VEGF and secrete approximately 2–3 pg of FGF-2 or VEGF throughout the experimental period exert a proangiogenic response when applied onto the CAM that is similar to the one elicited by 1 µg of recombinant cytokine (Fig. 1.10) (Ribatti et al., 2001).

Many techniques can be applied within the constraints of paraffin and plastic embedding, including histochemistry and immunohistochemistry. Electron microscopy can also be used in combination with light microscopy. Moreover, unfixed sponges can be utilized for chemical studies, such as the determination of DNA, protein, and collagen content, as well as for RT-PCR analysis of gene expression by infiltrating cells, including endothelial cells.

Shell-less culture of avian embryos facilitates experimental access and continuous observations of the growing embryos almost to the term of hatching. Hence it has a range of applications in developmental biology, angiogenesis, and pharmacology research. The embryo and its extraembryonic membranes may be transferred to a Petri dish on day 3 or 4 of incubation and CAM develops at the top as a flat

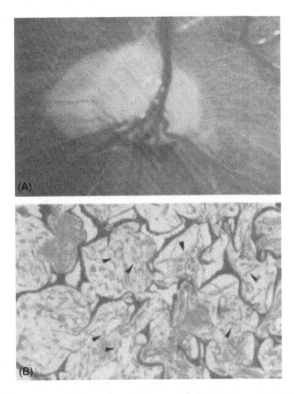

Figure 1.10 Effect of VEGF-overexpressing V-12-MCF-7 cells delivered on the top of the CAM using a gelatin sponge implant. (A) Macroscopic observation of the CAM showing the sponge surrounded by allantoic vessels that develop radially toward the implant in a "spoked-wheel" pattern. (B) Histological analysis showing a highly vascularized tissue among the sponge trabeculae (arrow heads). Reproduced from Ribatti, D., Nico, B., Morbidelli, L., et al., 2001. Cell-mediated delivery of fibroblast growth factor-2 and vascular endothelial growth factor onto the chick chorioallantoic membrane: endothelial fenestrations and angiogenesis. J. Vasc. Res. 38. 536−545.

membrane and reaches the edge of the dish to provide a 2D monolayer onto which multiple grafts can be placed (Auerbach et al., 1974). This system has several advantages as the accessibility of the embryo is greatly improved outside of the shell. Shell-less culture is much more amenable to live imaging than in vivo techniques. However, long-term viability is often lower in shell-less cultures and great attention must be paid to preventing the embryo from drying out. In the original description of embryos cultured in Petri dishes, there was a 50% loss in the first 3 days after cracking due to the frequent rupture of the yolk membrane, with 80% of those which survive to day 7 until day 16 (Auerbach et al., 1974). The ex vivo method is preferred to the in vivo method because it allows the quantification of the response over a wider area of the CAM.

EVALUATION OF THE VASOPROLIFERATIVE RESPONSE BY SEMIQUANTITATIVE METHODS

Several semiquantitative methods have been used to evaluate the extent of the vasoproliferative response. One method considers variations in the distribution and density of CAM vessels next to the site of grafting: these are evaluated in vivo by means of a stereomicroscope at regular intervals following the grafting procedure. The response is rated equal to 0 when no change with respect to the time of grafting can be appreciated; the score is +1 when few microvessels converging toward the implant are observed; +2 when a considerable change in the number and distribution of converging microvessels is appreciated (Knigton et al., 1977).

By another method, the degree of vasoproliferative response is defined as a vascular index based upon photographic reconstructions. All the vessels converging toward the implant and contained inside a 1 mm in diameter ring superposed to the CAM are enumerated: the ring is drawn around the implant in such a way that it will form an angle of less than 45 degrees with a straight line drawn starting from the implant's center. Vessels branching dichotomically outside the ring are counted as 2, while those branching inside the ring are counted as 1 (Fig. 1.11). (Barnhill and Ryan, 1983; Folkman and Cotran, 1976; Barrie et al., 1993). A third method expresses the degree of vasoproliferative response as evaluated in vivo under the stereomicroscope by

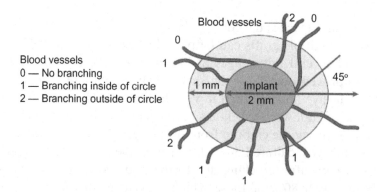

Figure 1.11 Evaluation of a proangiogenic response by macroscopic semiquantitative scoring of vessel branching. The drawing shows representative examples of different branching responses with scores branching with scores ranging from 0 to 2. Reproduced from Ribatti, D., Nico, B., Vacca, A., et al., 2006. The gelatin sponge-chorioallantoic membrane assays. Nat. Prot. 1, 85–91.

means of a 0–5 scale of arbitrary values. Zero describes a condition of the vascular network that is unchanged with respect to the time of grafting; 1 marks a slight increment in vessel density associated with occasional changes in the course of vessels converging toward the grafting site; 2, 3, 4, and 5 correspond to a gradual increase in vessel density associated with increased irregularity in their course. A coefficient describing the degree of angiogenesis can also be derived from the ratio of the calculated value to the highest attainable value; thus, the coefficient's lowest value is equal to 0 and the highest value is 1 (Fig. 1.12) (Folkman and Cotran, 1976).

When an angiostatic compound is tested, a semiquantitative evaluation of the antiangiogenic response is performed by two independent observers, which determine the radius of the growth inhibition zone as 0–4 grades vessel growth from the center of each disk soaked with the angiostatic compound to the furthest contiguous area in which tertiary vessels are absent. Zones with a radius greater than 1 mm are interpreted as evidence of significant inhibition of angiogenesis (Barrie et al., 1993).

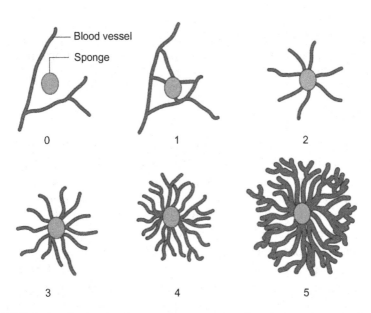

Figure 1.12 Evaluation of a proangiogenic response by macroscopic semiquantitative scoring ranging from 0 to 5. Reproduced from Ribatti, D., Nico, B., Vacca, A., et al., 2006. The gelatin sponge-chorioallantoic membrane assays. Nat. Prot. 1, 85–91.

EVALUATION OF THE VASOPROLIFERATIVE RESPONSE BY QUANTITATIVE METHODS

Quantitative evaluation of vascular density can be obtained by applying morphometric and planimetric methods to the observation of histologic CAM specimens. The angiogenic response can be evaluated as microvessel area by using a morphometric method of "point counting" (Ribatti et al., 1997). With a double-headed photomicroscope, two investigators simultaneously identify transversally cut microvessels (diameter ranging from 3 to 10 μm), and each identification is agreed upon in turn. Microvessels are studied at magnification of 250 times, with a square mesh inserted in the eyepiece. The microvessel area is indicated by the final mean number of the occupied intersection points expressed again as percentage of the total number of intersection points. Automatic image analyses have been suggested for CAM (Jakob et al., 1978). Nevertheless, Richardson and Singh (2003) report that "the approaches to the use of the CAM remain empirical, uncritical, and inconsistent, precluding meaningful comparison of data generated in different laboratories." To overcome these problems, digital cameras mounted on microscopes are used and the vessel density is calculated with the help of computer programs (Kirchner et al., 1996; Parsons-Wingerter et al., 1998).

More recently, Ejaz et al. (2016) have proposed a novel 3D model of the developing CAM for precise quantification of normal vasculature of the CAM from day 4 to day 13 of incubation.

GENOMICS

Genomics and bioinformatics tools are currently available for a variety of model systems, and avian embryology has begun to adopt these techniques. Comparative genomic revealed that the chick genome is three times smaller than the one of both human and mouse, but contains approximately the same number of genes (Burt, 2005). Modern advances, including the creation of the annotated chick genome (Wallis et al., 2004), gene expression profiling, live imaging, improved somatic transgenesis, and gene-specific attenuation of RNA levels, have also added to the classical strengths of the avian embryo. Seventy million bp of the chicken sequence is highly conserved with humans both within coding gene segments and outside coding regions (ICGSC, 2004). The complete characterization of the chick embryo genome (www.nhgri.gov/11510730) will be helpful to synthesize a broad panel of antibodies with high specificity for chicken tissues, for blood and lymphatic endothelial cells and stroma components. This aspect could be useful to better characterize the interactions between implanted human and/or mouse tumors and chicken tissues.

Gene expression profiling associated with the physiological CAM development as well as with the angiogenic switch during tumor progression has been reported (Javerzat et al., 2009; Saidi et al., 2008). Engraftment of human tumor tissue onto the CAM, followed by transcriptomic analyses with both human and chicken microarrays, enables the gene signatures of both the host stroma and the human tumor to be distinguished. Soulet et al. (2010) performed wounding of the chicken CAM and compared gene expression to normal CAM at the same stage of development. Bioinformatics analysis leads to the identification of several new genes with an endothelial cell signature. Moreover, Soulet et al. (2013) applied in vivo biotinylation combined with high-resolution mass spectrometry and bioinformatic analyses to study the vascular and matrix proteome in the CAM. Exertier et al. (2014) performed a Affymetrix gene screening for VEGF-A-induced genes during CAM vascularization. A total of 53 single genes with human orthologs and preferential endothelial expression were identified, and this list contained numerous key angiogenic regulators with known endothelial expression.

Retroviral, lentiviral, and adenoviral vectors have been used to infect the CAM, leading to the expression of the viral transgene enabling the monitoring of small subpopulations of transgenic cells within a tissue (Hen et al., 2012). Moreover, this allows the long-lasting presence of the gene product that is expressed directly by CAM cells, and makes feasible the study of the effects of intracellular or membrane-bound proteins as well as of dominant-negative gene products. This approach will shed new lights on the study of the metastatic process by using the CAM assay.

Advantages
1. High embryo survival rate.
2. Easy methodology.
3. Low cost.
4. Reproducibility.

5. Reliability.
6. The CAM does not require administrative procedures for obtaining ethics committee approval for animal experimentation, because the chick embryo is not considered as living animal until day 17 of development in most countries. The CAM is not innervated, and experiments are terminated before the development of centers in the brain associated with pain perception, making this a system not requiring animal experimentation permissions. The Institutional Animal Care and Use Committee (IACUC), an Association of New England Medical Center and Tufts (IACUC, 2001), as well as the National Institutes of Health, United States (National Institute of Health, 1991), established that a chick embryo that has not reached the 14th day of its gestation period would not experience pain and can therefore be used for experimentation without any ethical restrictions or prior protocol approval.
7. The chick embryo cannot mount an "immune" response to foreign tumor cells until well after day 12, but it can respond to tumor cells by infiltration of monocytes and inflammatory-like cells such as avian heterophils. Being naturally immunodeficient, the chick embryo may receive transplantations from different tissues and species, without immune responses.

Disadvantages
1. The main limitation of CAM is represented by nonspecific inflammatory reactions which may develop as a result of grafting, and in turn induce a secondary vasoproliferative response eventually making it difficult to quantify the primary response that is being investigated (Jakob et al., 1978; Spanel-Burowski et al., 1988). Inflammatory angiogenesis per se in which infiltrating macrophages or other leukocytes may be the source of angiogenic factors cannot be distinguished from direct angiogenic activity of the test material without detailed histological study and multiple positive and negative controls. In this context, a study of histological CAM section would help detecting the possible presence of a perivascular inflammatory infiltrate together with a hyperplastic reaction, if any, of the chorionic epithelium. However, the possibilities of causing nonspecific inflammatory response are much lower when the test material is grafted as soon as CAM begins to develop since then the host's immune system is relatively immature (Leene et al., 1973). This problem may be overcome by using the yolk sac vessels of the

4-day chick embryo because this system has a markedly reduced inflammatory and immune response (Taylor and Weiss, 1984; Ribatti, 1995).

2. False-positive angiogenesis may be induced by any test material which causes cell damage by virtue of abnormal osmolarity, pH, or toxicity. Such substrates may induce an inflammatory response or cause focal contraction of the CAM.

3. Angiogenesis may be induced by degradation product of fibrin (Thompson et al., 1985) which can leak from CAM vessels in response to injurious test substances.

4. The test material is placed on preexisting vessels and it derives that the actual neovascularization can hardly be distinguished from a falsely increased vascular density due to a rearrangement of preexisting vessels that follows contraction of the membrane (Knighton et al., 1991).

5. Timing of the CAM angiogenic response is essential. Many studies determine angiogenesis after 24 h, a time at which there is no angiogenesis, but only vasodilation. It would be worthwhile to point out that measurements of vessel density are really measurement of new-formed vessels, and that the distinction between vasodilation and neovascularization is not easy to make. To circumvent this drawback it is useful to utilize sequential photography to document new vessel formation.

6. The CAM is also extremely sensitive to modification by environmental factors, such as changes in oxygen tension, which make the sealing of the opening in the shell critical, pH, osmolarity, and the amount of keratinization.

REFERENCES

Auerbach, R., Kubai, L., Knighton, D.R., et al., 1974. A simple procedure for the long-term cultivation of chicken embryos. Dev. Biol. 41, 391–394.

Ausprunk, D.H., Knighton, D.R., Folkman, J., 1974. Differentiation of vascular endothelium in the chick chorioallantois: a structural and autoradiographic study. Dev. Biol. 38, 237–249.

Ausprunk, D.H., Knighton, D.R., Folkman, J., 1975. Vascularization of normal and neoplastic tissues grafted to the chick chorioallantois. Am. J. Pathol. 79, 597–610.

Barnhill, R.L., Ryan, T.J., 1983. Biochemical modulation of angiogenesis in the chorioallantoic membrane of the chick embryo. J. Invest. Dermatol. 81, 485–488.

Barrie, R., Woltering, E.A., Hajarizadeh, H., et al., 1993. Inhibition of angiogenesis by somatostatin and somatostatin-like compounds is structurally dependent. J. Surg. Res. 55, 446–450.

Burt, D.W., 2005. Chicken genome: current status and future opportunities. Genome Res. 15, 1692–1698.

Chambers, A.F., Schmidt, E.E., MacDonald, I.C., et al., 1992. Early steps in hematogenous metastasis of B16F1 melanoma cells in chick embryos studied by high-resolution intravital video-microscopy. J. Natl. Cancer Inst. 84, 797–803.

Dugan, J.D.J., Lawton, M.T., Glaser, B., et al., 1991. A new technique for explanation and in vitro cultivation of chicken embryos. Anat. Rec. 229, 125–128.

Dunn, B., Fitzharris, T., Barnett, B., 1981. Effects of varying chamber construction and embryo pre-incubation age on survival and growth of chick embryo in shell-less culture. Anat. Rec. 199, 33–43.

Durupt, F., Koppers-Lalic, D., Balme, B., et al., 2012. The chicken chorioallantoic membrane tumor assay as model for qualitative testing of oncolytic adenoviruses. Cancer Gene. Ther. 19, 58–68.

Ejaz, S., Chekarova, I., Ashraf, M., et al., 2016. A novel 3D model of chick chorioallantoic membrane for ameliorated studies in angiogenesis. Cancer Invest. 24, 567–575.

Exertier, P., Javerzat, S., Wang, B., et al., 2014. Impaired angiogenesis and tumor development by inhibition of the mitotic kinesin Eg5. Oncotarget 4, 2302–2316.

Folkman, J., Langer, R., Linhardt, R., et al., 1983. Angiogenesis inhibition and tumor regression caused by heparin or a heparin fragment tin the presence of cortisone. Science 221, 719–725.

Fuchs, A., Lindenbaum, E.S., 1988. The two- and three-dimensional structure of the microcirculation of the chick chorioallantoic membrane. Acta Anat. 131, 271–275.

Hagedorn, M., Javerzat, S., Gilges, D., et al., 2005. Accessing key steps of human tumor progression in vivo by using an avian embryo model. Proc. Natl. Acad. Sci. U.S.A. 102, 1643–1648.

Gualandris, A., Rusnati, M., Belleri, M., et al., 1996. Basic fibroblast growth factor overexpression in endothelial cells: an autocrine mechanism for angiogenesis and angioproliferative diseases. Cell Growth Diff. 7, 147–160.

Guidolin, D., Nico, B., Mazzocchi, G., et al., 2004. Order and disorder in the vascular network. Leukemia 18, 1745–1750.

Hen, G., Yosefi, S., Shinder, D., et al., 2012. Gene transfer to chicks using lentiviral vectors administered via the embryonic chorioallantoic membrane. PLoS One 7, e36531.

IACUC, 2001. Policy on Protocol Approval for Use of Chicken Embryos and Eggs. An Association of New England Medical Center and Tufts.

ICGSC, 2004. Sequence and comparative analysis of the chicken genome provide unique perspectives on vertebrate evolution. Nature 432, 695–716.

Javerzat, S., Franco, M., Herbert, J., et al., 2009. Correlating global gene regulation to angiogenesis in the developing chick extra-embryonic vascular system. PLoS One 4, e7856.

Khokha, R., Zimmer, M.J., Wilson, S.M., et al., 1992. Up-regulation of TIMP-1 expression in B16-F10 melanoma cells suppresses their metastatic ability in chick embryo. Clin. Exp. Metast. 10, 365–370.

Kim, J., Yu, W., Kovalski, K., et al., 1998. Requirement for specific proteases in cancer cell intravasation as revealed by a novel semiquantitative PCR-based assay. Cell 94, 353–362.

Knighton, D.R., Ausprunk, D.H., Tapper, D., et al., 1977. Avascular and vascular phases of tumor growth in the chick embryo. Br. J. Cancer 35, 347–356.

Knighton, D.R., Fiegel, V.D., Philipps, G.D., 1991. The assays for angiogenesis. Progr. Clin. Biol. Res. 365, 291–299.

Koop, S., Khokha, R., Schmidt, E.E., et al., 1994. Overexpression of metalloproteinase inhibitor in B16F10 cells does not affect extravasation but reduces tumor growth. Cancer Res. 54, 4791–4797.

Koop, S., Schmidt, E.E., MacDonald, I.C., et al., 1996. Independence of metastatic ability and extravasation: metastatic ras-transformed and control fibroblasts extravasate equally well. Proc. Natl. Acad. Sci. U.S.A. 93, 11080–11084.

Kirchner, L.M., Schmidt, S.P., Gruber, B.S., 1996. Quantitation of angiogenesis in the chick chorioallantoic membrane model using fractal analysis. Microvasc. Res. 51, 2–14.

Jakob, W., Jentzsch, K.D., Manersberger, B., et al., 1978. The chick chorioallantoic membrane as bioassay for angiogenesis factors: reactions induced by carrier materials. Exp. Pathol. 15, 241–249.

Jakobson, A., Hahnenberger, R., Magnusson, A., 1989. A simple method for shell-less cultivation of chick embryos. Pharmacol. Toxicol. 64, 193–195.

Jankovic, B.D., Isakovic, K., Lukic, M.L., et al., 1975. Immunological capacity of the chicken embryo. I. Relationship between the maturation of lymphoid tissues and the occurrence of cell-mediated immunity in the developing chicken embryo. Immunology 29, 497–508.

Janse, E.M., Jeurissen, S.H., 1991. Ontogeny and function of two non-lymphoid cell populations in the chicken embryo. Immunobiol. 182, 472–481.

Langer, R., Folkman, J., 1976. Polymers for the sustained release of proteins and other macromolecules. Nature 363, 475–482.

Leene, W., Duyzings, M.J.M., VonSteeg, C., 1973. Lymphoid stem cell identification in the developing thymus and bursa of Fabricius of the chick. Z. Zellforsh. 136, 521–533.

Lugassy, C., Barnhill, R.L., 2007. Angiotropic melanoma and extravascular migratory metastasis: a review. Adv. Anat. Pathol. 14, 195–201.

Lugassy, C., Vernon, S.E., Busam, K., et al., 2006. Angiotropism of human melanoma: studies involving in transit and other cutaneous metastases and the chicken chorioallantoic membrane: implications for extravascular melanoma invasion and metastasis. Am. J. Dermatopathol. 28, 187–193.

MacDonald, I.C., Schmidt, E.E., Morris, V.L., et al., 1992. Intravital videomicroscopy of the chorioallantoic microcirculation: a model system for studying metastasis. Microvasc. Res. 44, 185–199.

Mira, E., Lacalle, R.A., Gomez-Mouton, C., et al., 2002. Quantitative determination of tumor cell intravasation in a real-time polymerase chain reaction-based assay. Clin. Exp. Metast. 19, 313–318.

Nguyen, M., Shing, Y., Folkman, J., 1994. Quantitation of angiogenesis and antiangiogenesis in the chick embryo chorioallantoic membrane. Microvasc. Res. 47, 31–40.

Ossowski, L., Reich, E., 1980. Experimental model for quantitative study of metastasis. Cancer Res. 40, 2300–2309.

Parsons-Wingerter, P., Lwai, B., Yang, M.C., et al., 1998. A novel assay of angiogenesis in the quail chorioallantoic membrane: stimulation by bFGF and inhibition by angiostatin according to fractal dimension and grid intersection. Microvasc. Res. 55, 201–214.

Ribatti, D., 1995. A morphometric study of the expansion of the chick area vasculosa in shell-less culture. J. Anat. 186, 639–644.

Ribatti, D., Roncali, L., Nico, B., et al., 1987. Effects of exogenous heparin on vasculogenesis of the chorioallantoic membrane. Acta. Anat. 130, 257–263.

Ribatti, D., Vacca, A., Bertossi, M., et al., 1990. Angiogenesis induced by B-cell non-Hodgkin's lymphomas. Lack of correlation with tumor malignancy and immunologic phenotype. Anticancer Res. 10, 401–406.

Ribatti, D., Urbinati, C., Nico, B., et al., 1995. Endogenous basic fibroblast growth factor is implicated in the vascularization of the chick embryo chorioallantoic membrane. Dev. Biol. 170, 39–49.

Ribatti, D., Vacca, A., Roncali, L., et al., 1996. The chick embryo chorioallantoic membrane as a model for in vivo research on angiogenesis. Int. J. Dev. Biol. 40, 1189–1197.

Ribatti, D., Gualandris, A., Bastaki, M., et al., 1997. New model for the study of angiogenesis and antiangiogenesis in the chick embryo chorioallantoic membrane: the gelatin sponge/chorioal-lantoic membrane assay. J. Vasc. Res. 34, 455–463.

Ribatti, D., Alessandri, G., Vacca, A., et al., 1998. Human neuroblastoma cells produce extracel-lular matrix-degrading enzymes, induce endothelial cell proliferation and are angiogenic in vivo. Int. J. Cancer 77, 449–454.

Ribatti, D., Nico, B., Morbidelli, L., et al., 2001. Cell-mediated delivery of fibroblast growth factor-2 and vascular endothelial growth factor onto the chick chorioallantoic membrane: endo-thelial fenestrations and angiogenesis. J. Vasc. Res. 38, 536–545.

Ribatti, D., Nico, B., Vacca, A., et al., 2006. The gelatin sponge-chorioallantoic membrane assays. Nat. Prot. 1, 85–91.

Richardson, M., Singh, G., 2003. Observations on the use of the avian chorioallantoic membrane (CAM) model in investigations into angiogenesis. Current Drug Targets Cardiovasc. Haematol. Disord. 3, 155–185.

Saidi, A., Javerzat, S., Bellahcene, A., et al., 2008. Experimental anti-angiogenesis causes upregu-lation of genes associated with poor survival in glioblastoma. Int. J. Cancer 122, 2187–2198.

Schlatter, P., Konig, M.F., Karlsson, L.M., et al., 1997. Quantitative study of intussusceptive capillary growth in the chorioallantoic membrane (CAM) of the chicken embryo. Microvasc. Res. 54, 65–73.

Spanel-Burowski, K., Schnapper, U., Heymer, B., 1988. The chick chorioallantoic membrane assay in the assessment of angiogenic factors. Biomed. Res. 9, 253–260.

Soulet, F., Kilarski, W.W., Antczak, P., et al., 2010. Gene signatures in wound tissue as evi-denced by molecular profiling in the chick embryo model. BMC Genom. 11, 495.

Soulet, F., Kilarski, W.W., Roux-Dalvai, F., et al., 2013. Mapping the extracellular and mem-brane proteome associated with the vasculature and the stroma in the embryo. Mol. Cell Proteom. 12, 2293–2312.

Taylor, C.M., Weiss, C.B., 1984. The chick vitelline membrane as a new test system for angiogen-esis and antiangiogenesis. Int. J. Microcirc. Clin. Exp. 3, 337.

Taylor, S., Folkman, J., 1982. Protamine is an inhibitor of angiogenesis. Nature 297, 307–312.

Thompson, W.D., Campbell, R., Evans, T., 1985. Fibrin degradation response in the chick embryo chorioallantoic membrane. J. Pathol. 145, 27–37.

Vacca, A., Ribatti, D., Iurlaro, M., et al., 1998. Human lymphoblastoid cells produce extracellu-lar matrix-degrading enzymes and induce endothelial cell proliferation, migration, morphogenesis, and angiogenesis. Int. J. Clin. Lab. Res. 28, 55–68.

Vacca, A., Ribatti, D., Presta, M., et al., 1999. Bone marrow neovascularization, plasma cell angiogenic potential, and matrix metalloproteinase-2 secretion parallel progression of human mul-tiple myeloma. Blood 93, 3064–3073.

van der Horst, E.H., Leupold, J.H., Schubbert, R., et al., 2004. TaqMan-based quantification of invasive cells in the chick embryo metastasis assay. Biotechniques 37, 940–942.

Yu, W., Kim, J., Ossowski, L., 1997. Reduction in surface urokinase receptor forces malignant cells into a protracted state of dormancy. J. Cell Biol. 137, 767–777.

Weber, W.T., Mausner, R., 1977. Migration patterns of avian embryonic bone marrow cells and their differentiation to functional T and B cells. Adv. Exp. Med. Biol. 88, 47–59.

Wilting, J., Christ, B., Bokeloh, M., 1991. A modified chorioallantoic membrane (CAM) assay for qualitative and quantitative study of growth factors. Studies on the effects of carriers, PBS, angiogenin and bFGF. Anat. Embryol. 183, 259–271.

FURTHER READING

Dusseau, J.W., Hutchins, P.M., Malsaba, D.S., 1986. Stimulation of angiogenesis by adenosine on the chick chorioallantoic membrane. Circ. Res. 59, 163–170.

CHAPTER 2

The Corneal Assay for Angiogenesis

RABBIT CORNEAL ASSAY

The transparent, avascular human cornea has a diameter of about 12 mm and is approximately 0.5 mm thick. The cornea consists of five layers: the surface epithelium, Bowman's membrane, the stroma which forms the major part of the cornea, Descemet's membrane, and the endothelium (Fig. 2.1). The corneal epithelium has a rich nerve supply and it consists of two to three cell layers of flattened superficial cells, two to three cell layers of wing cells, and a single layer of columnar basal cells. Beneath the corneal epithelium lie the Bowman's membrane, a resistant, and an acellular collagen structure. The corneal stroma forms 85%–90% of the thickness of the entire cornea. It is made of regularly arranged collagen fibrils, which are responsible for corneal transparency. The collagen matrix contains keratocytes, fibroblast-like cells which produce substances essential for the maintenance of the homeostasis of the cornea. The noncellular Descemet's membrane, secreted by the cells of the corneal endothelium, is located between the stroma and the internal endothelium. The corneal endothelium consists of a single layer of polygonal, flattened cells. Their main role is to extract water from the stroma so that the arrangement of the collagen matrix remains regular. Corneal transparency has been found to be dependent on many factors: the rapid renewal of the epithelium, the maintenance of the integrity of its structure, the state of relative dehydration of the stroma, the absence of blood vessels, the normal metabolic activity of keratocytes, and the cells of corneal endothelium, which have a vital role in the maintenance of the transparency and the normal function of the cornea.

A number of mechanisms have been identified in the cornea to explain its avascularity, including (1) densely packed collagen lamellae and the presence of compact collagen networks resulting in a mechanical defense; (2) angiostatic nature of corneal epithelial cells; (3) induction of anterior chamber associated immune deviation, in which antigen-specific delayed-type hypersensitivity is suppressed;

In Vivo Models to Study Angiogenesis. DOI: http://dx.doi.org/10.1016/B978-0-12-814020-8.00002-0

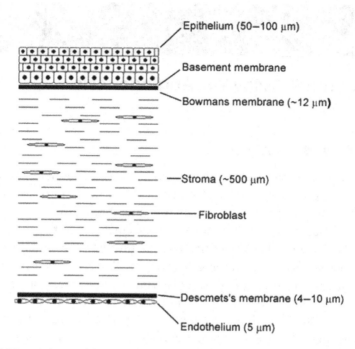

Figure 2.1 Schematic drawing of the structure of the cornea.

(4) extensive innervation; (5) low levels of angiogenic factors; (6) the barrier function of limbal cells; and (7) active production of potent antiangiogenic factors (Ambati et al., 2006; Azar, 2006; Cursiefen et al., 2006). An important aspect of corneal avascularity is closely linked to integrity of the corneal epithelium.

Unlike the corneal endothelium, the corneal epithelium has an extensive capacity of regeneration upon injury. A functionally intact epithelium is another prerequisite of corneal avascularity.

As early as the 1870s, the Dutch ophthalmologist van Dooremaal noted prolonged survival of mouse skin graft placed in the anterior chamber of a dog eye (Chong and Dana, 2008). It is known that the cornea and anterior chamber are immunologically quiescent by a complicated, but not fully understood, active process (Azar, 2006). Niederkorn (2003) described the immunological privilege in terms of a three-legged stool where each leg represents one arm of the corneal immune privilege. The first leg is the afferent (entry) blockade, constituting avascularity and low expression of major histocompatibility complex II (MHC class II) (antigen presentation) on corneal cells.

The second leg is the central processing, where immune deviation is induced, diverting the immune reaction from a cytotoxic to a relatively more benign response. Finally, the third leg constitutes the blockade of the efferent arm, where soluble and membrane-bound factors induce apoptosis of immune cells, inhibiting them from migrating to nearby lymph nodes to consequently induce an immune reaction (Niederkorn, 2003; Chong and Dana, 2008).

The corneal assay consists in the placement of an angiogenesis inducer (tumor tissue, cell suspension, growth factor) into a micro-pocket produced in the cornea thickness in order to evoke vascular outgrowth from the peripherally located limbal vasculature (Fig. 2.2). This assay, with respect to the other in vivo assays, has the advantage of measuring only new blood vessels, since the cornea is initially avascular.

The rabbit cornea has been used since 1953 to demonstrate inflam-matory neovascularization (Ashatan and Cook, 1953). The corneal assay performed in New Zealand white rabbits 4–6 months old was firstly described by Gimbrone et al. (1974). As Judah Folkman remem-bered "Michael Gimbrone, a postdoctoral fellow, and I implanted tumors (of approximately 0.5 mm^3) into the stromal layers of the rab-bit cornea at distances of up 2 mm from the limbal edge. New capillary blood vessels grew from the limbus, invaded the stroma of the avascu-lar corneas, and reached the edge of the tumor over a period of approximately 8–10 days. When tumors were implanted beyond 3 mm

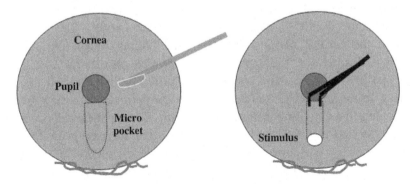

Figure 2.2 Schematic representation of the corneal micropocket assay. A micropocket is surgically produced in the corneal stroma of anaesthetized animals by a surgical scalpel and a pliable spatula (A) and the test substance is inserted in the micropocket (B). Reproduced from Morbidelli, L., 2003. The corneal assay for angiogenesis, In: Ribatti, D., Vacca, A., Dammacco, F. (Eds.), In Vitro and In Vivo Models of Angiogenesis. Servizio Editoriale Universitario, Bari, pp. 9–19.

from the limbus (or in the center of rabbit cornea, which is approximately 12 mm in diameter), no neovascularization was observed" (Folkman, 2008). The tumor implants were usually 1 mm in size and tumor growth and neovascular response of limbal vessels were studied by means of a slit-lamp stereomicroscope, histological examination, colloidal carbon injections, and autoradiography after exposure to ^{3}H-labeled thymidine (Gimbrone et al., 1974).

One of the first demonstrations that tumors could induce neovascularization, as opposed to merely inducing vessel dilation, was performed by introducing tumor pieces into the aqueous humor of the anterior chamber of the rabbit eye, remote from the iris (Gimbrone et al., 1972). Tumors remained viable, avascular, limited in size (<1 mm^3), and induce neovascularization of iris vessels, but are too remote from these vessels to invade them (Gimbrone et al., 1972). The tumor floats and stays dormant.

This experiment introduced the concept of tumor dormancy used in the sense of "population dormancy." In a population of tumor cells not penetrated by new capillaries, there is an outer proliferating compartment balanced by dying cells in a central necrotic compartment. Tumors grown in the vitreous of the rabbit eye remain viable, but attain diameters of less than 0.50 mm for as long as 100 days. Once such a tumor reaches the retinal surface, it becomes neovascularized and within 2 weeks can undergo a 19.000-fold increase in volume over the avascular tumor (Brem et al., 1976).

Other assays directly implanted tissue fragments on the iris (Gimbrone et al., 1973). The iris transplant starts with a 2 mm incision through the cornea, about 1 mm from the limbus. When the aqueous humor is flushed out, the explants are deposited on the iris. If a tumor (e.g., the Brown-Pierce carcinoma) is implanted onto the rabbit iris, i.e., a site with an angiogenic potential, it is soon permeated by new vessels, its proliferative activity becomes exponential, all neoplastic cells continue to multiply and very few are lost from the system, its mass grows rapidly, and the steady state is reached at very high levels (4000−16,000 times the original volume) (Fig. 2.3) (Gimbrone et al., 1972).

Ziche and Gulino (1982) by using the rabbit cornea demonstrated that angiogenesis is a marker of neoplastic transformation. Normal mouse diploid fibroblasts were carried in culture. At each passage,

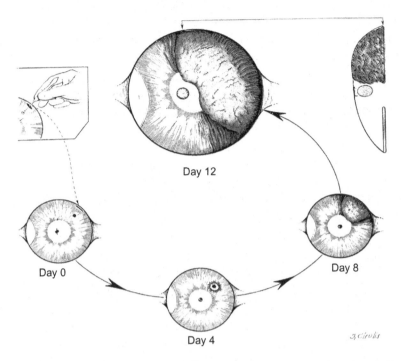

Day 12

Day 0

Day 8

Day 4

Figure 2.3 The patterns of development of two simultaneous implants of Brown–Pierce tumor in the rabbit eye. The anterior chamber implant remains avascular and stops expanding at a small size, while the iris implant vascularizes and grows progressively. Reproduced from Gimbrone, M.A., Leapman, S., Cotran, R.S., et al., 1972. Tumor dormancy in vivo by prevention of neovascularization. J. Exp. Med. 136, 261–267.

the cells were tested for angiogenic activity in the rabbit eye, and for tumorigenicity by reimplantation into the mouse strain that donates the fibroblasts. Angiogenesis first appeared at the fifth passage, while tumorigenicity did not occur until the fifteenth passage.

This assay was subsequently refined by introducing the tumor cells into the stroma of the cornea, and then by substituting slow-release pellets containing known quantities of semipurified angiogenic growth factors from tumor cells. Nonspecific angiogenic stimuli such as endotoxin were used initially but were replaced with specific growth factors such as fibroblast growth factor-2 (FGF-2) and vascular endothelial growth factor (VEGF).

After initial looping of limbal capillaries, loops extend into the cornea and vascular sprouts appear at the apices of the loops, and finally a vascular network develops toward the tumor implant.

The rabbit size (2–3 kg) lets an easy manipulation of both the whole animal and the eye to be easily extruded from its location and to be surgically manipulated. Sodium pentobarbital is used to anaesthetize animals and under aseptic conditions a micropocket (1.5 mm × 3 mm) is produced using a pliable iris spatula 1.5 mm in width in the lower half of the cornea. A small amount of the aqueous humor can be drained from the anterior chamber when reduced corneal tension is required. For a corneal transplant, a 1.5 mm incision is made just off center of the corneal dome to a depth of about one-half the thickness of the cornea. Substances used within the pocket have included tumor tissues, tumor cells, tumor cell extracts, other tissues and cells, concentrated conditioned medium, purified recombinant cytokines, and growth factors. The implant is located at 2.5–3 mm from the limbus to avoid false positives due to mechanical procedure and to allow the diffusion of test substances in the tissue, with the formation of a gradient for the endothelial cells of the limbal vessels (Fig. 2.4). Implants sequestering the test material and the control are coded and implanted in a double masked manner. The material under test can be in the form of slow-release pellets incorporating recombinant growth factors, cell suspensions, or tissue samples. When a fluid is to be tested, the material can be injected directly into the cornea.

During the first 48 h after the transplant, slight opacity of the cornea usually appears. Corneal edema seems to precede vascularization and a loosening of the corneal stroma facilitates the invasion of the tissue by blood vessels. If the opacity does not disappear by the third day, it is the expression of an inflammatory reaction.

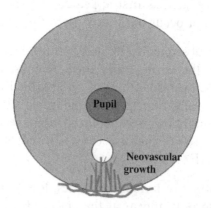

Figure 2.4 The newly formed vessels start from the limbal vasculature and progress toward the implanted stimulus. Reproduced from Morbidelli, L., 2003. The corneal assay for angiogenesis, In: Ribatti, D., Vacca, A., Dammacco, F. (Eds.), In Vitro and In Vivo Models of Angiogenesis. Servizio Editoriale Universitario, Bari, pp. 9–19.

Recombinant growth factors are prepared as slow-release pellets by incorporating the substance under test into an ethylene vinyl acetate copolymer (Elvax-40) or poly-2-hydroxethyl methacrylate (Hydron) (Langer and Folkman, 1976). Elvax-40 is not biodegradable within the body; it is quite inert and causes little or no reaction following implantation. Before use, it is important to wash the polymer in absolute ethanol at 37°C for 10–15 days until the pellet fails to induce any inflammatory reaction. Hydron is idrophobic and when the polymer is subjected to water, it will swell due to the molecule's hydrophilic pendent group. For such implants it has possible to release proteins and other macromolecules at nearly constant rates of micrograms/day and nanograms/day for a period of weeks to months. Rhine et al. (1980) proposed a theoretical mechanism for this new technique of controlled release of macromolecules.

Several angiogenic factors in combination can be simultaneously implanted into the corneal tissue and synergistic or antagonistic angiogenic response can be analyzed. Moreover, inhibition of corneal neovascularization can be achieved by systemic administration of angiogenesis inhibitors. One of the first inhibitors of angiogenesis studied by means of this assay was an extract of cartilage (Brem and Folkman, 1975). V2 carcinoma and neonatal cartilage were implanted together in a corneal pocket. When cartilage was implanted with tumor the vessels were inhibited from reaching the tumor, while with inactive cartilage the vessels entered the tumor and rapid tumor growth follows.

Cortisone and hydrocortisone inhibit corneal vascularization in man, especially after subconjunctival administration and have an inhibitory effect on experimentally produced corneal vascularization produced by thermal or chemical burns and alloxan.

The corneal assay may also be used to study lymphangiogenesis. Different lymphangiogenic factors can be implanted into the corneal tissue and the newly formed lymphatic vessels can be detected using specific markers, including LYVE-1, podoplanin, vascular endothelial growth factor receptors (VEGFRs) and Prox-1 (Fig. 2.5) (Cao et al., 2006; Björndahl et al., 2005a). Moreover, implantation of tumors into the cornea allows to study associated tumor lymphangiogenesis (Björndahl et al., 2005b).

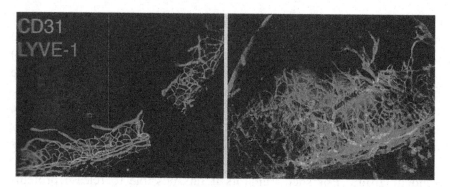

Figure 2.5 CD31-positive blood vessels and LYVE-1-positive lymphatic vessels in the mouse cornea. On the left, preexisting vessels; on the right, newly formed vessels in response to a cytokine stimulation. Reproduced from Nakao, S., Haferi-Maghadam, A., 2016. The corneal micropocket assay: a model for angiogenesis and lymphangiogenesis. Method. Mol. Biol. 1430, 311–316.

When the overexpression of growth factors by stable transfection of specific cDNA is studied, one eye is implanted with transfected cells and the other with the wild-type cell line (Ziche et al., 1997). When tissue samples are tested, samples of 2–3 mg are obtained by cutting the fresh tissue fragments under sterile conditions, and the angiogenic activity of tumor samples is compared with macroscopically healthy tissue (Gallo et al., 1998). When drug solutions incompatible with Elvax polymerization and genes transduced by viral vectors have to be tested, after the removal of aqueous humor, a volume of 10 μL is injected within the corneal stroma in the space between the limbus and the pellet implant.

Regression of blood vessels (pruning) is an essential process during vessel maturation in embryonic and postnatal development. Hyaloid vessels of the eye are a prime example for a vessel type that is temporally formed during embryonic development but regresses at later stages of development. Although it is still not fully clear whether vessel regression is regulated by an active genetic program or whether it is simply due to withdrawal of prosurvival angiogenic factors that usually maintain the vasculature, pruning is associated with apoptotic endothelial cells. Corneal blood vessels can regress after the removal of the angiogenic stimulus (Ausprunk et al., 1976) (Fig. 2.6) photodynamic (Bucher et al., 2014) or antiangiogenic therapy (Koenig et al., 2012), whereas corneal lymphatic vessels form and regress spontaneously within the first weeks after birth (Zhang et al., 2011).

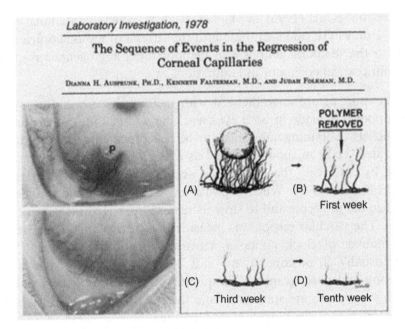

Figure 2.6 The first report of the sequential morphological events that occur during regression of neovascularization after removal of an angiogenic stimulus. Reproduced from Ausprunk, D.H., Falterman, K., Folkman, J., 1976. The sequence of events in the regression of corneal capillaries. Lab. Invest. 38, 284–294.

There is still considerable variation in angiogenic and lymphangiogenic response depending on the genetic background of the mouse strain. This shows that while corneal pocket assays are a good method for initial screening of proangiogenic substances, care should be taken when quantitatively comparing the results.

MOUSE CORNEAL ASSAY

This assay was firstly described by Muthukkaruppan and Auerbach (1979). The mouse corneal micropocket angiogenesis assay has capitalized upon this property and is often considered the gold standard to determine whether signaling peptides, drugs, etc. function as proangiogenic or antiangiogenic factors in vivo. Muthukkaruppan and Auerbach (1979) also examined the angiogenic response resulting from implantation of immunocompetent lymph node fragments. They observed vasodilation in the limbal region adjacent to the graft, followed by the extension of new vascular loops in the cornea. However, lymphocyte-induced angiogenesis is not associated with the formation of capillary sprouts from the apices of capillary loops.

Kenyon et al. (1996) implanted Hydron pellets containing either FGF-2 or VEGF and sucralfate into the stroma of mouse cornea adjacent to the temporal limbus and demonstrated angiogenesis response without corneal edema or inflammation.

The animals are anaesthetized with methosyflurane, and a corneal micropocket is made in both eyes reaching within 1 mm of the limbus and pellets containing substances to be tested coated with Hydron are implanted. Hydron should be used as a casting solution, and was prepared by dissolving the polymer in absolute alcohol at 37°C (Langer and Folkman, 1976). When peptides are tested, sucralfate is added to stabilize the molecule and to slow its release from Hydron (Chen et al., 1995). The vascular response is measured as the maximal vessel length and number of clock hours of neovascularization is scored at fixed time (usually on postoperative 5 and 7 days) using a slit-lamp biomicroscopy and photographed. To quantify the section of the cornea in which new vessels are sprouting from the preexisting limbal vessels, the circumference of the cornea is divided into the equivalent of 12 clock hours, and the measure of the number of clock hours of neovascularization for each eye is performed during each observation.

RAT CORNEAL ASSAY

Purified growth factors are combined 1:1 with Hydron (Polverini and Leibovich, 1984). Pellets are implanted 1−1.5 mm from the limbus of the cornea of anaesthetized rats. Neovascularization is assessed at 3, 5, and 7 days; animals are perfused with colloidal carbon solution to label vessels; eyes are enucleated and fixed in 10% neutral buffered formalin overnight. The following day, corneas are excised, flattened, and photographed. A positive neovascularization response is recorded only if sustained directional ingrowth of capillary sprouts and hairpin loops toward the implant is observed. Negative responses are recorded when either no growth is observed or when only an occasional sprout or hairpin loop showing no evidence of sustained growth is detected.

Treatment With Drugs

The effect of local drug treatment on corneal neovascularization can be studied in the form of ocular drops or ointment (Presta et al., 1999) or microinjection in the corneal thickness. The effect of systemic drug treatment on corneal angiogenesis can also be evaluated

(Ziche et al., 1997). However, when considering the size of the animals, systemic drug treatment in rabbits requires a higher amount of drugs than smaller animals. The use of nude mice allows the study of angiogenesis modulation in response to effectors produced and released by tumors or tumor cell lines of human origin growing subcutaneously. The corneal assay has been further modified and adapted in severe combined immunodeficiency mice, to detect circulating inhibitors of angiogenesis generated by human tumors grown in immunodeficient mice (Chen et al., 1995).

Several genetically engineered mice developing spontaneous corneal neovascularization have been developed, and transgenic models of corneal neovascularization can complement retinal and choroidal neovascularization models and help to clarify mechanisms of de novo angiogenesis (Table 2.1). In transgenic corneal neovascularization

Table 2.1 Transgenic Mouse Models	
Model	Mechanism
Destrin$^{-/-}$ mice, corn1 mice	Proangiogenic VEGFR-3 signaling, deficient in sVEGFR-1
TSP-1$^{-/-}$ mice	Loss of antiangiogenic signaling of CD36$^+$ macrophages
CD36$^{-/-}$ mice	Loss of TSP-1−CD36 interaction
KLEIP/KLHL20$^{-/-}$ mice	Proangiogenic miR-204-Ang-1 pathway
LRIG1$^{-/-}$ mice	LRIG1 acts as a negative regulator of STAT3
K5.Stat3C mice	Proinflammatory STAT3 signaling
Jam-A$^{-/-}$ mice	VEGF expression and TGF-β activation
ADAMTS9$^{+/-}$ mice	ADAMTS9 is an antiangiogenic metalloprotease
LeCre;Cited2$^{loxP/loxP}$	Presumably via Pax6 and Klf4
MxCre;RBP$^{f/f}$	Downstream of Notch receptor
Pax6 overexpression using the Aldh3a1 promoter	Chi3l4, Flt-1, Wif1
Pax6$^{+/-}$ mice	Presumably due to epithelial defects; sVEGFR-1 deficiency
Global FoxC1$^{-/-}$ mice and neural-crest-specific FoxC1$^{-/-}$	Via VEGF−VEGFR-2 signaling
LeCre;VEGFR-2$^{loxP/loxP}$	Lack of sVEGFR-2
pCre;VEGFR-1$^{loxP/loxP}$	Lack of sVEGFR-1
C57BL/6 mice	Unknown
Nude mice, nu/nu, and hairless mutant mouse strain, SKH1;h/h	Proangiogenic factors probably coming from the epithelium
Heme oxygenase (HO)-2	MMP-2

models, proangiogenic signaling is partly due to canonical VEGF signaling (like the destrin$^{-/-}$/corn1 model) or due to VEGF-independent signaling (like in late stage KLEIP$^{-/-}$corneal neovascularization models). In several transgenic corneal neovascularization models, an involvement of several different proangiogenic mechanisms is conceivable.

Quantification of the Angiogenic Response

An angiogenic response is scored positive when budding of vessels from the limbal plexus occurs after 3−4 days and capillaries progress to reach the implanted pellet in 7−10 days. Implants that fail to produce a neovascular growth within 10 days are considered negative, while implants showing an inflammatory reaction are discarded (Ziche et al., 1989). The potency of angiogenic activity is evaluated on the basis of the number and growth rate of newly formed capillaries, and an angiogenic score is calculated by the formula [vessel density × distance from limbus] (Ziche et al., 1997). A density value of 1 corresponds to 0 to 25 vessels per cornea, 2 from 25 to 50, 3 from 50 to 75, 4 from 75 to 100, and 5 for more than 100 vessels. The distance from the limbus is graded with the aid of an ocular grid. To understand the mechanism of progression/regression consequent to treatment, vessel density and length are considered separately, documenting the activity of treatment on endothelial cell proliferation (density) compared to elongation and organization (length).

It is also possible to use computerized image analysis. Pictures are taken from each eye, and images are digitalized and analyzed by an ad hoc software after the extraction of the newly formed vessels form the background, providing a more objective assessment of the neovascular response a period of time (Fig. 2.7) (Proia et al., 1988). The total number of vessels, the area occupied by vessels, and the branching of the neovascular net are measured and statistically analyzed. Intravenously administered fluorochrome-labeled high-molecular-weight dextran (FITC dextran) has been employed for visualization of corneal neoangiogenesis (Kenyon et al., 1996).

Histology and Scanning Electron Microscopy

Newly formed vessels and the presence of inflammatory cells are detected by hematoxylin/eosin staining or specific immunohistochemical procedure, i.e., anti-rabbit macrophages (RAM11), anti-CD-31 for

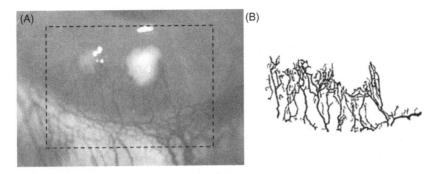

Figure 2.7 Computerized image analysis of the angiogenic response induced by the implant of two adjacent pellets releasing two angiogenic stimuli (A). Following digitalization, the new vessels are extracted from the background (B). The image, converted in black and white, is analyzed for the number of vessels, the area occupied by vessels, and the degree of branching. Reproduced from Morbidelli, L., 2003. The corneal assay for angiogenesis, In: Ribatti, D., Vacca, A., Dammacco, F. (Eds.), In Vitro and In Vivo Models of Angiogenesis. Servizio Editoriale Universitario, Bari, pp. 9–19.

endothelium (Ziche et al., 1997). Double staining (i.e., anti-CD-31 for vascular endothelium and specific markers for tumor cells) could be useful to label newly formed vessels of the host and proliferating tumor cells implanted in the cornea.

Neovascularization may be studied by scanning electron microscopy of vascular casts (Burger et al., 1983). The initial response is a vasodilation of pericorneal vessels, followed by the formation of new capillary buds from the pericorneal venules and capillaries. These sprouts lengthen and multiply to produce a rich anastomosing plexus. Vascular casts emphasize capillaries and venules as the predominant source of new vessels. Moreover, ultrastructural evidence suggests that migration and redistribution of existing endothelial cells from the limbal vessels enables vascular sprouting and elongation without cellular proliferation (Sholley et al., 1984).

Advantages

1. New vessels are easily identified.
2. The ability to monitor progress of angiogenesis in the absence of an existing background vasculature. This simplifies the quantification of the neovascular area, thus removing a source of variation. It also eliminates the possibility of vessel dilation being mistaken for angiogenesis.
3. The cornea is an immunologically privileged site before vascularization. Indeed, because histological examination of all immunemodified

and control corneas was consistent for the presence of only minimal inflammatory cellular activity, it seems unlikely that inflammation constitutes a major stimulus to the observed angiogenesis.

4. Rabbits are more docile and amenable to handling and experimentation than mice and rats.

Disadvantages

1. Rabbit cornea has been found avascular in all strains examined so far. In some strains of rats the presence of preexisting vessels within the cornea and the development of keratitis are serious disadvantages. Due to the three-dimensional nature of vessel growth, it is difficult to quantify the vascular response except in the early stages, when the vessel length and number can be monitored.

2. The degree of inflammation may influence the degree and duration of neovascularization. In case of inflammatory reactions, these are easily detectable in rabbits by stereomicroscopic examination as corneal opacity. Nevertheless, the corneal model has provided a powerful tool to study inflammation-induced angiogenesis through implantation of any solid material or chemical or mechanical injury (Vu et al., 1985).

3. In mice and rats it is possible to obtain time point results. The small size of the murine eye makes the procedure even more technically demanding and requires skilled operators. The evolution of the angiogenic response in the same animal is not recommended because each time the cornea is observed the animal has to be anaesthetized since it is not easy to keep it quiet. Experiments are made with a large number of animals and vessel growth during time can be visualized by perfusion with colloidal carbon solution in individual animals. Multiple observations are instead possible in rabbits. The use of slit-lamp stereomicroscope and of not anaesthetized animals allows the observation of newly formed vessels during time with long time monitoring, even for 1–2 months.

4. The technique is expansive, time consuming, and technically demanding, so that relatively few animals can be grafted at a single setting. Although it is less expensive to use rats or mice than to use rabbits, surgery becomes more difficult as the size of the eye gets smaller.

5. The space for introducing test material is limited. Due to the size of the eye, there is an inherent limitation on the size of the implanted tissue construct. It is important to note that cornea is not the only

mouse transplantation site that is amenable to live imaging of tissue constructs. The cranial window and dorsal skinfold have been successfully employed in live imaging studies. Although these methods are better suited for the implantation of larger tissue constructs and constitute more "typical" cellular environments as compared to the highly specialized, avascular cornea, the surgical procedures are fairly laborious for the researcher and invasive to the animal.

REFERENCES

Ambati, B.K., Nozaki, M., Singh, N., et al., 2006. Corneal avascularity is due to soluble VEGF receptor-1. Nature 443, 993–997.

Ashatan, N., Cook, C., 1953. Mechanisms of corneal neovascularization. Br. J. Ophtalmol. 37, 193–209.

Ausprunk, D.H., Falterman, K., Folkman, J., 1976. The sequence of events in the regression of corneal capillaries. Lab. Invest. 38, 284–294.

Azar, D.T., 2006. Corneal angiogenic privilege: angiogenic and antiangiogenic factors in corneal avascularity, vasculogenesis, and wound healing. Trans. Am. Ophthalmol. Soc. 104, 264–302.

Björndahl, M.A., Cao, R., Nissen, L.J., et al., 2005a. Insulin-like growth factor-1 and -2 induce lymphangiogenesis in vivo. Proc. Natl. Acad. Sci. U.S.A. 102, 5593–5598.

Björndahl, M.A., Cao, R., Burton, J.B., et al., 2005b. Vascular endothelial growth factor-A promotes peritumoral lymphangiogenesis and lymphatic metastasis. Cancer Res. 65, 9261–9266.

Brem, S., Brem, H., Folkman, J., et al., 1976. Prolonged tumor dormancy by prevention of neovascularization in the vitreous. Cancer Res. 36, 2807–2812.

Bucher, F., Bi, Y., Gehlsen, U., et al., 2014. Expression of mature lymphatic vessels in the cornea by photodynamic therapy. Br. J. Ophthalmol. 98, 391–395.

Burger, P.C., Chandler, D.B., Klintworth, G.K., 1983. Corneal neovascularisation as studied by scanning electron microscopy of vascular casts. Lab. Invest. 48, 169–180.

Cao, R., Björndahl, M.A., Gallego, M.I., et al., 2006. Hepatocyte growth factor is a lymphangiogenic factor with an indirect mechanism of action. Blood 107, 3531–3536.

Chen, C., Parangi, S., Tolentino, M.T., et al., 1995. A strategy to discover circulating angiogenesis inhibitors generated by human tumors. Cancer Res. 55, 4230–4233.

Chong, E.M., Dana, R.M., 2008. Graft failure IV. Immunologic mechanisms of corneal transplant rejection. Int. Ophthalmol. 28, 209–222.

Cursiefen, C., Chen, L., Saint-Geniez, M., et al., 2006. Nonvascular VEGF receptor 3 expression by corneal epithelium maintains avascularity and vision. Proc. Natl. Acad. Sci. U.S.A. 30, 11405–11410.

Folkman, J., 2008. Tumor angiogenesis from bench to bedside. In: Marmé, D., Fusenig, N. (Eds.), Tumor Angiogenesis. Springer, Berlin, pp. 3-2, 2008.

Gallo, O., Masini, E., Morbidelli, L., et al., 1998. Role of nitric oxide in angiogenesis and tumor progression in head and neck cancer. J. Natl. Cancer Inst. 90, 587–596.

Gimbrone, M.A., Leapman, S., Cotran, R.S., et al., 1972. Tumor dormancy in vivo by prevention of neovascularization. J. Exp. Med. 136, 261–267.

Gimbrone, M.A., Leapman, S., Cotran, R.S., et al., 1973. Tumor angiogenesis: iris neovascularisation at a distance from experimental intraocular tumors. J Natl. Cancer Inst. 50, 219–228.

Gimbrone, M.A., Cotran, R., Leapman, S.B., et al., 1974. Tumor growth and neovascularization: an experimental model using the rabbit cornea. J. Natl. Cancer Inst. 52, 413–427.

Kenyon, B.M., Voest, E.E., Chen, C.C., et al., 1996. A model of angiogenesis in the mouse cornea Invest. Ophthalmol. Vis. Sci. 37, 1625–1632.

Koenig, Y., Bock, F., Kruse, F.E., et al., 2012. Angioregressive pretreatment of mature corneal blood vessels before keratoplasty: fine-needle vessel coagulation combined with anti-VEGFs. Cornea 31, 887–892.

Langer, R., Folkman, J., 1976. Polymers for the sustained release of proteins and other macromolecules. Nature 363, 797–800.

Morbidelli, L., 2003. The corneal assay for angiogenesis. In: Ribatti, D., Vacca, A., Dammacco, F. (Eds.), In Vitro and In Vivo Models of Angiogenesis. Servizio Editoriale Universitario, Bari, pp. 9–19.

Muthukkaruppan, V., Auerbach, R., 1979. Angiogenesis in the mouse cornea. Science 206, 1416–1418.

Nakao, S., Haferi-Maghadam, A., 2016. The corneal micropocket assay: a model for angiogenesis and lymphangiogenesis. Method. Mol. Biol. 1430, 311–316.

Niederkorn, J.Y., 2003. The immune privilege of corneal grafts. J. Leukoc. Biol. 74, 167–171.

Polverini, P.J., Leibovich, S.J., 1984. Induction of neovascularization in vivo and endothelial cell proliferation in vitro by tumor associated macrophages. Lab. Invest. 51, 635–642.

Presta, M., Rusnati, M., Belleri, M., et al., 1999. Purine analog 6-methylmercaptopurine ribose inhibits early and late phases of the angiogenesis process. Cancer Res. 59, 2417–2424.

Proia, A.D., Chandler, D.B., Haynes, W.L., et al., 1988. Quantitation of corneal neovascularization using computerized image analysis. Lab. Invest. 58, 473–479.

Sholley, M.M., Ferguson, G.P., Seibel, H.R., et al., 1984. Mechanisms of neovascularisation. Vascular sprouting can occur without proliferation of endothelial cells. , Lab. Invest. 51, 624–634.

Vu, M.T., Burger, P.C., Klinthworth, G.K., 1985. Angiogenic activity in injured rat corneas as assayed in the chick chorioallantoic membrane. Lab. Invest. 53, 311–319.

Zhang, H., Hu, X., Tse, J., et al., 2011. Spontaneous lymphatic vessel formation and regression in the murine cornea. Invest. Ophthalmol. Vis. Sci. 52, 334–338.

Ziche, M., Gullino, P.M., 1982. Angiogenesis and neoplastic progression in vitro. J. Natl. Cancer Inst. 69, 483–487.

Ziche, M., Alessandri, G., Gullino, P.M., 1989. Gangliosides promote the angiogenic response. Lab. Invest. 61, 629–634.

Ziche, M., Morbidelli, L., Choudhuri, R., et al., 1997. Nitric oxide-synthase lies downstream of vascular endothelial growth factor but not basic fibroblast growth factor induced angiogenesis. J. Clin. Invest. 99, 2625–2634.

FURTHER READING

Voest, E.E., Kenyon, B.M., O'Relly, M.S., et al., 1995. Inhibition of angiogenesis in vivo by interleukin 12. J. Natl. Cancer Inst. 87, 581–586.

Ziche, M., Jones, J., Gullino, P.M., 1982. Role of prostaglandinE1 and copper in angiogenesis. J. Natl. Cancer Inst. 69, 475–482.

Ziche, M., Morbidelli, L., Masini, E., et al., 1994. Nitric oxide mediates angiogenesis in vivo and endothelial cell growth and migration in vitro promoted by substance P. J. Clin. Invest. 94, 2036–2044.

CHAPTER 3

The Matrigel Implant Assay

MATRIGEL ASSAY

The most subcutaneous implant models were developed with the aim to trap a putative angiogenic compound into a suitable carrier, mostly an avascular sponge-like structure, which can slowly release the factor at the site of implant. The final result of this process is the recruitment of new blood vessels. These tests can also be used to study the in vivo activity of antiangiogenic agents. Passaniti et al. (1992) described a simple adaptation of the cutaneous implant assays using reconstituted basement membranes. Reconstituted basement membranes have been extensively used to study specific steps of the angiogenic process in vitro and assess angiogenesis in vivo (Benelli and Albini, 1999).

Three-dimensional matrix substrates have been extensively used in vitro to study the differentiation of endothelial cells into vessel-like structures: fibrin and collagen gels have proven good substrates (Montesano et al., 1983); however, the extensive time required for these assays is a major limiting factor. These problems can be bypassed by using a reconstituted basement membrane, the most widely used material is known as Matrigel, an extract of a murine tumor which produces massive amounts of basement membrane (Kleinman et al., 1986).

Matrigel is a mixture of basement membrane components prepared from the murine Engelbreth−Holm−Swarm (EHS) tumor of C57/ black mice grown in vivo. It contains collagen IV, laminin, nidogen/ entactin, heparan sulfate proteoglycan, and growth factors including epidermal growth factor (EGF), transforming growth factor beta (TGF-β), platelet derived growth factor (PDGF), nerve growth factor (NGF), fibroblast growth factor-2 (FGF-2), and insulin like growth factor (ILGF) (Vukicevic et al. 1992). Moreover, Matrigel also contained perclan, a heparan sulfate proteoglycan of the basement membrane, which is involved in the storage of angiogenic growth factors (Aviezer et al., 1997; Sharma et al., 1998). Matrigel is liquid at 4°C and forms a gel when warmed at 37°C (Kleinman et al., 1982).

In Vivo Models to Study Angiogenesis. DOI: http://dx.doi.org/10.1016/B978-0-12-814020-8.00003-2

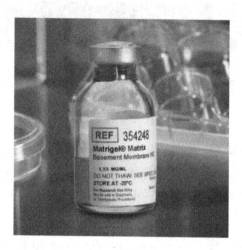

Figure 3.1 A sample of commercial Matrigel.

Commercially prepared Matrigel is generally supplied in aliquots at 10–12 mg/mL.

A commercial growth factor–reduced Matrigel is also available (Fig. 3.1) in which the levels of stimulatory cytokines and growth factors have been markedly reduced. Moreover, this preparation is obtained by reducing the content of heparan sulfate proteoglycan (low-molecular-mass proteins are soluble in 20% of saturated ammonium sulfate unlike laminin, collagen IV, and heparan sulfate), thus altering Matrigel composition.

Matrigel can be frozen and thawed several times without any problem; it is convenient to prepare small aliquots to shorten the thawing procedure in ice/water bath. As Matrigel quickly polymerizes at room temperature, all the material used for its manipulation (pipettes, tips, tubes, multiwells) must be prechilled. Matrigel is often used in three different tests related to angiogenesis: the chemoinvasion assay, the morphology assay, and the in vivo sponge model (Fig. 3.2). An improvement to the assay is the encapsulation of the Matrigel in a Plexiglas chamber or within flexible plastic tubing prior to implantation (Krag et al., 2003).

Matrigel can also be used for morphological studies on vascular cell organization. Matrigel is thawed at 4°C in an ice/water bath, and 0.3–0.4 mL of a concentrated solution (10 mg/mL) was carefully pipetted into 13-mm diameter tissue culture wells (48-well chambers), thus avoiding even small bubbles. Matrigel is then polymerized for 1 h

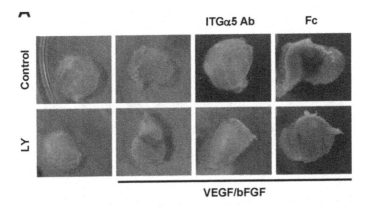

Figure 3.2 LY-2157299 induces angiogenesis in vivo in a Matrigel-plug assay and promotes VEGF + FGF-2-induced angiogenesis. Matrigel plugs combined with PBS (control), VEGF and FGF-2, and/or LY-2157299, in the presence or absence of α5-integrin-neutralizing antibody or Fc control protein, were subcutaneously injected into mice. LY-2157299 induces angiogenesis and enhances VEGF + FGF-2-induced angiogenesis. The synergistic effect of LY + VEGF + FGF-2 on angiogenesis is inhibited by an anti-α5-integrin antibody but not by Fc. Reproduced from Liu, Z., Kobayashi, K., van Dinther, M., et al 2009. VEGF and inhibitors of TGFbeta type-I receptor kinase synergistically promote blood-vessel formation by inducing alpha5-integrin expression. J Cell Sci. 122,3294−302 .

at 37°C. Once polymerization is occurred, 7×10^4 endothelial cells in 1 mL of medium are carefully pipetted on top of the gel. The plates are then incubated at 37°C in a 5% CO_2, humidified atmosphere. The assays are photographed and monitored with an inverted microscope, a modification using tetrazolium dye (MTT) colorimetric assay as a dye has been recently described (Sasaki and Passaniti, 1998). Human umbilical vein endothelial cells suspended in 10% fetal calf serum (FCS)-containing medium with heparan and endothelial cell growth supplement (ECGS) organize into capillary-like networks within 6 h (Kubota et al., 1998).

This assay has been used in the preclinical evaluation of angiogenesis inhibitors (Taraboletti et al., 1995). In this case, treatment is usually begun early, when vessels are not yet formed, and continues throughout the whole duration of the experiment. In addition, the Matrigel plug assay can be used to study vascular-targeting agents (Micheletti et al., 2003). Treatment with these agents, which act on existing neovessels, is done when vessels are already formed (5−7 days after Matrigel implantation). Shortly before autopsy, mice receive an intravenous injection of fluorescent tracers such as fluorescein isothiocyanate-conjugated Griffonia simplicifolia isolectin B4 and the collected pellets are then analyzed under confocal microscopy, where the three-dimensional structure of permeated, fluorescent vessels can be used as an index of vessel functionality.

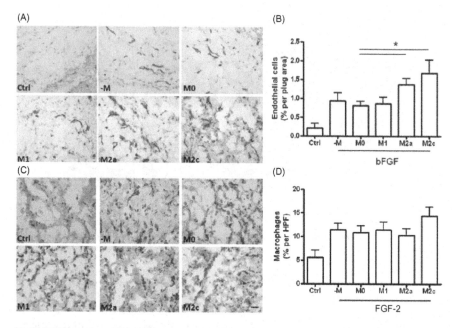

Figure 3.3 Effect of macrophage subsets on angiogenesis in vivo. CD31 to stain endothelial cells (A) and anti-monocyte-macrophage-2 (MOMA-2) to stain macrophages (C) in Matrigel plugs after 14 days. (B) M2 macrophages increase the number of endothelial cells. (D) Macrophage subsets do not affect macrophage chemotaxis. Reproduced from Jetten, N., 2014. Macrophage heterogeneity in neovascularization. Uitgeverij BOXPress, Hertogenbosch.

Angiogenesis is estimated by embedding and sectioning plugs in paraffin and staining using Masson's trichrome, which stain Matrigel blue and endothelial cells red. The endothelial cells are also positive to factor VIII and CD31 (Fig. 3.3) (Passaniti et al., 1992). As index of angiogenesis it is also possible to measure hemoglobin level with the Drabkin assay, but it is not possible to distinguish between blood in capillaries and blood in the sinuses of larger vessels (Passaniti et al., 1992). Vessel formation in young mice (6-month-old) is reduced compared to older mice. Moreover, lower angiogenic response is observed if the material is injected into the dorsal surface of the animal, while one of the best areas in terms of angiogenic response is the ventral surface of the mouse in the grain area close to the dorsal midline. Alternative methods involve quantification of the vasculature after the injection of fluorochrome-labeled, high-molecular-weight dextran and the quantitative assessment of the vasculature in a chamber model (Baker et al., 2006).

CHEMOINVASION ASSAY

As Albini and Benelli (2008) pointed out: "The chemoinvasion assay can give affordable and reproducible results once established for the cell lines of interest." This is a modification of the classic Boyden chemotaxis assay in Boyden chambers (Fig. 3.4) (Albini et al., 1987). Polyvinylpyrrolidone (PVP)-free polycarbonate filters are coated with Matrigel diluted with cold water on ice. Filters are placed on tissue culture plates and the liquid Matrigel is pipetted onto the filter. During this step care must be taken to insure that the Matrigel solution is homogeneously applied on filter surface. The Matrigel-coated filters are dried under a laminar flow and the plate tightly closed with parafilm. The filters are rehydrated just before performing the assay using serum-free Dulbecco's modified Eagle's medium (DMEM) at 4°C. Numerous substances can be placed as chemoattractants in the lower chamber. Kaposi's sarcoma cell conditioned medium is a good chemoattractant for endothelial cells, mimicking a highly angiogenic environment (Albini et al., 1992); purified angiogenic growth factors are

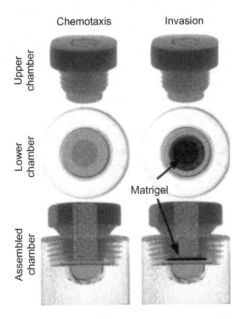

Figure 3.4 The use of blind wells in chemotaxis and chemoinvasion: the only difference between a chemotaxis (left) and an invasion chamber (right) is the addition of a Matrigel coating to chemotaxis filters (cyan, light gray in print versions). The Matrigel-coated area (pink, black in print versions) must be larger than the surface of the lower chamber of the blind well containing the chemoattractants (orange, dark gray in print versions). Reproduced from Albini, A., Benelli, R., 2008. The chemoinvasion assay: a method to assess tumor and endothelial cell invasion and its modulation. Nat. Prot. 3, 504–511.

also frequently used, i.e., VEGF or FGF-2. Both chemoattractants and cells are suspended in serum-free medium containing 0.1% BSA (SFM). Medium with BSA alone in the lower chamber serves as a negative control. Chemotaxis and chemoinvasion assays are often done in parallel to determine if treatments affect chemotaxis itself or if they are specific for invasion. A control of chemotaxis in a chemoinvasion assay is also useful to verify that Matrigel concentration is sufficient to act as a real barrier to cell migration. An invasion index (invasion/chemotaxis) can be calculated and indicate the specific contribution of matrix degradation. The upper chamber is filled with an endothelial cell suspension (usually 120.000 cells/800 μL/chamber). The chambers are then incubated at 37°C in 5% CO_2 for 6 h. While tumor cells usually do not migrate in response to SFM alone, endothelial cells show a variable degree of random unstimulated migration due to the loss of cell—cell contact; once determined, this value is usually subtracted to the final data as "background migration." Invasion inhibitors acting on the chemoattractant are added to the lower compartment of the Boyden chamber, while inhibitors acting on the cells or cell products are generally added along with the cells. At the end of the incubation time, cells remaining on the upper surface of the filter are mechanically removed by wiping them with a cotton swab or stripping on a glass slide. The cells migrated to the undersurface are quantitated after staining. In the original assay quantitation was performed by microscope counting five to ten random fields for each filter. Various alternatives have been proposed such as colorimetric detection of the staining, image analysis, and metabolic labeling with MTT or similar compounds (Albini, 1998).

Advantages
1. Technically simple. Matrigel is readily implanted by a simple injection into the ventral region of mice where it solidifies forming a plug and avoiding and complex surgical procedures.
2. Suitable for large-scale screening.
3. Rapid quantitative analysis.
4. As Matrigel plug is initially avascular, any vessel found in the plug reflects a neovascularization process.

Disadvantages
1. Costly.
2. Analysis is time-consuming.

3. The presence of growth factors and cytokines as a component of Matrigel render difficulties in the extrapolation of the results.

4. The implants become encapsulated and elicit the formation of foreign body giant cells which secrete angiogenic cytokines, thereby interfering with the test substances.

5. The assay suffers from considerable variability, because it is difficult to generate identical three-dimensional plugs, even though the total Matrigel volume is kept constant.

REFERENCES

Albini, A., Iwamoto, Y., Kleinman, H.K., et al., 1987. A rapid in vitro assay for quantitating the invasive potential of tumor cells. Cancer Res. 47, 3239–3245.

Albini, A., Repetto, L., Carlone, S., et al., 1992. Characterization of Kaposi's sarcoma-derived cell cultures from an epidemic and a classic case. Int. J. Oncol. 1, 723–730.

Albini, A., 1998. Tumor and endothelial cell invasion of basement membranes. Pathol. Oncol. Res. 4, 1–12.

Albini, A., Benelli, R., 2008. The chemoinvasion assay: a method to assess tumor and endothelial cell invasion and its modulation. Nat. Prot. 3, 504–511.

Aviezer, D., Iozzo, R.V., Noonan, D.M., et al., 1997. Suppression of autocrine and paracrine functions of basic fibroblast growth factor by stable expression of perlecan antisense cDNA. Mol. Cell. Biol. 17, 1938–1946.

Baker, J.H.E., Huxman, I.A., Kyle, A.H., et al., 2006. Vascular-specific quantification in an in vivo Matrigel chamber angiogenic assay. Microvasc. Res. 71, 69–75.

Benelli, R., Albini, A., 1999. In vitro models of angiogenesis: the use of Matrigel. Int. J. Biol. Markers 14, 243–246.

Jetten, N., 2014. Macrophage Heterogeneity in Neovascularization. Uitgeverij BOXPress, Hertogenbosch.

Kleinman, H.K., McGarvey, M.L., Hassell, J.R., et al., 1986. Basement membrane complexes with biological activity. Biochemistry 25, 312–318.

Kleinman, H.K., Mc Gravy, M.L., Liotta, L.A., et al., 1982. Isolation and characterization of type IV procollagen, laminin, and heparan sulfate proteoglycan from EHS sarcoma. Biochemistry 21, 6188–6193.

Krag, M., Hjarnaa, P.J.V., Bramm, E., et al., 2003. In vivo chamber angiogenesis assay: an optimized Matrigel plug assay for fast assessment of anti-angiogenic activity. Int. J. Oncol. 22, 305–311.

Kubota, Y., kleinman, H.K., Martin, G.R., et al., 1998. Role of laminin and basement membrane in the morphological differentiation of human endothelial cells into capillary-like structures. J. Cell. Biol. 107, 1589–1598.

Liu, Z., Kobayashi, K., van Dinther, M., et al., 2009. VEGF and inhibitors of TGFbeta type-I receptor kinase synergistically promote blood-vessel formation by inducing alpha5-integrin expression. J. Cell Sci. 122, 3294–3302.

Micheletti, G., Poli, M., Borsotti, P., et al., 2003. Vascular-targeting activity of ZD6126, a novel tubulin-binding agent. Cancer Res. 63, 1534–1537.

Montesano, R., Orci, L., Vassalli, P., 1983. In vitro rapid organization of endothelial cells into capillary-like networks is promoted by collagen matrices. J. Cell. Biol. 97, 1648–1652.

Passaniti, A., Taylor, R.M., Pili, R., et al., 1992. A simple, quantitative method for assessing angiogenesis and antiangiogenic agents using reconstituted basement membrane, heparin, and fibroblast growth factor. Lab. Invest. 67, 519–528.

Sasaki, C., Passaniti, A., 1998. Identification of anti-invasive but non-cytotoxic chemotherapeutic agents using the tetrazolium dye MTT to quantitate viable cells in Matrigel. Biotechniques 24, 1038–1043.

Sharma, B., Handler, M., Eichstetter, I., et al., 1998. Antisense targeting of perlecan blocks tumor growth and angiogenesis in vivo. J. Clin. Invest. 102, 1599–1608.

Taraboletti, G., Garofalo, A., Belotti, D., et al., 1995. Inhibition of angiogenesis and murine hemangioma growth by batimastat, a synthetic inhibitor of matrix metalloproteinases. J. Natl. Cancer Inst. 87, 293–298.

Vukicevic, S., Kleinman, H.K., Luyten, F.P., et al., 1992. Identification of multiple active growth factors in basement membrane Matrigel suggest caution in interpretation of cellular activity related to extracellular matrix components. Exp. Cell. Res. 202, 1–8.

CHAPTER 4

The Sponge Implant Model

THE EXPERIMENTAL MODEL

The technique involves the implantation of sterile sponge disks into subcutaneous pockets of the dorsum of mice, and compounds of interest are injected directly into the sponge (Andrade et al., 1987a). Polyvinyl sponge implants were firstly used in dogs, rats, and rabbits as a framework for the ingrowth of vascularized connective tissue and measurement of enzyme activities in newly formed tissue (Grindlay and Waugh, 1951; Woessner and Boucek, 1959; Edwards et al., 1960).

A number of different sponge matrices have been used, including polyvinyl alcohol, cellulose acetate, polyester, polyether, polyurethane, and gelatin sponge alone or in combination (Fig. 4.1). The sponge implant induces an inflammatory reaction so that a granulomatous reaction develops around and inside the sponge. Newly formed microvessels develop from the preexisting vessels and invade the sponge as early as 5 days postimplantation (Fig. 4.2) (Andrade et al., 1987a). The sponges are retrieved after 1−2 weeks and neovascularization can be quantified by histology and immunohistochemistry (Fig. 4.3). Surrogate markers can be used, such as measurement of hemoglobin (Pesenti et al., 1992) or DNA content (Buckley et al., 1985).

The cannulated sponge model has been adapted for mice (Mahadevan et al., 1989). Cannulated sponge disk of polyether polyurethane are most frequently used (Fig. 4.4). The disks are implanted on mouse dorsal skin. The cannula is esteriorized and substances to be tested (putative pro- or antiangiogenic factors) are injected directly into the sponge disk via the cannula. The sequential development of blood flow in the implant sponges, originally avascular, can be determined by measuring the rate of ^{133}Xe injected into the implants, to measure blood flow changes over a period of weeks, and monitor the vascular changes indirectly (Kety, 1949). As the sponge originally

In Vivo Models to Study Angiogenesis. DOI: http://dx.doi.org/10.1016/B978-0-12-814020-8.00004-4

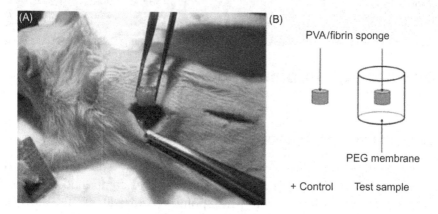

Figure 4.1 Subcutaneous implant model. (A) Placement of the gels into the subcutaneous pockets of the dorsal skin and (B)implant scheme: the fibrin-soaked poly-vinyl acetate (PVA) sponge is covered by a three-dimensional polyethylene glycol hydrogel shell and implanted as test construct. PVA sponges without membrane coverage were transplanted as positive controls. Reproduced from Wechsler, S., Fehr, D., Molenberg, A., et al., 2008. A novel, tissue occlusive poly(ethylene glycol) hydrogel material. J. Biomed. Mat. Res. Part A, 85, 285–292 (Wechsler et al., 2008).

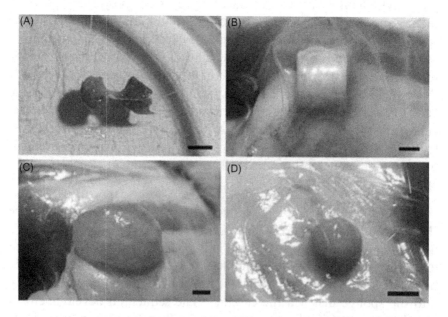

Figure 4.2 Explants at different time points. (A) Positive control 1.0 month after implantation. The sponge is well infiltrated with newly formed tissue; membrane samples (B) 1.0 month, (C) 4.3 months, and (D) 7.0 months postimplantation. Reproduced from Wechsler, S., Fehr, D., Molenberg, A., et al., 2008. A novel, tissue occlusive poly(ethylene glycol) hydrogel material. J. Biomed. Mat. Res. Part A, 85, 285–292.

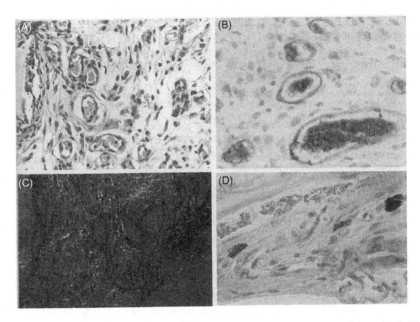

Figure 4.3 Histological sections of sponge implant. Hematoxylin and eosin (A); blood vessels stained with CD31 (B); picrosirius red staining for collagen (C); and Dominici Blue for mast cells (D). Reproduced from Andrade, S.P., Ferreira, M.A., 2016. The sponge implant model for angiogenesis. Method. Mol. Biol. 1430, 333–343 (Andrade and Ferreira, 2016).

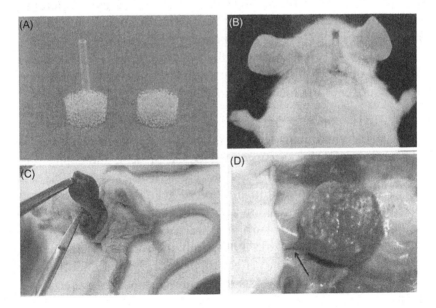

Figure 4.4 Implant sponge disk with and without cannula (A), and the subcutaneous arrangement of the implant in a mouse (B). Intraperitoneal implantation of synthetic matrix in mouse showing an adhesion-like tissue (C). A vascularized sponge 14 days postimplantation (D). Reproduced from Andrade, S.P., Ferreira, M.A., 2016. The sponge implant model for angiogenesis. Method. Mol. Biol. 1430, 333–343.

contained no blood vessels, the increase in the rate of ^{133}Xe loss from the sponges was considered an expression of neovascularization.

Low-molecular fluorochrome-complexed tracers have provided additional methods for detecting new blood vessels (Mc Grath et al., 1996). The measurement of fluorochrome-generated emission in the bloodstream following application in the sponge implant at various intervals postimplantation reflects the degree of local blood flow development and interaction of the angiogenic site with the systemic circulation. Measurement of blood flow indicates the functional state of the neovasculature (Andrade et al., 1987b).

The vascularization of the implants can be assessed by measuring the amount of hemoglobin contained in the tissues using the Drabkin method. Quantification of various biochemical parameters can be performed, including collagen metabolism (Paulini et al., 1974), fibronectin deposition (Holund et al., 1982), and proteoglycan turnover (Bollet et al., 1958). The technique also allows to characterize the sequence of histological changes in granulation tissue formation (Holund et al., 1979) and to estimate the kinetics of cellular proliferation (Davidson et al., 1985). The extent of neutrophils and macrophages infiltration in the sponges can be evaluated by estimating the enzymes myeloperoxidase and N-acetylglucosaminidase (Belo et al., 2004; Ferreira et al., 2004). The sponge model has been used to evaluate the angiogenic response in inflammatory tissue (Andrade et al., 1987b), to implant different tumor cell lines in rodents for studying tumor angiogenesis (Thiede et al., 1988; Andrade et al., 1992).

Polyvinyl alcohol sponges have been implanted subcutaneously in rats to study the angiogenic properties of a cartilage-derived growth factor and of EGF by measuring the DNA, protein, and collagen content of the implants (Davidson et al., 1985; Buckley et al., 1985). Two weeks after subcutaneous implantation of gelatin sponges, the sponges are removed and prepared for histology. By using this model, it has been demonstrated that suramin prevents neovascularization and tumor growth (Pesenti et al., 1992), and the role of thymidine phosphorylase in the angiogenic effect of platelet derived endothelial cell growth factor (PDECGF) has been investigated (Miyadera et al., 1995).

Advantage
1. The cannulated sponge model provides an objective and continuous assessment of neovascularization and is easily reproducible.

Disadvantages
1. In vivo monitoring of the angiogenic response is not possible except in the cannulated sponge and evaluation is restricted to fixed time points.
2. It induces nonspecific inflammation, so that the subcutaneous tissue encapsulates the sponge in a granulomatous reaction.
3. Differential sponge properties can affect angiogenesis quantification.

REFERENCES

Andrade, S.P., Fan, T.P., Lewis, G.P., 1987a. Quantitative in vivo studies on angiogenesis in a rat sponge model. Br. J. Exp. Pathol. 68, 755–766.

Andrade, S.P., Ferreira, M.A., 2016. The sponge implant model for angiogenesis. Method. Mol. Biol. 1430, 333–343.

Andrade, S.P., Machado, R.D., Teixeira, A., et al., 1987b. Sponge-induced angiogenesis in mice and the pharmacological reactivity of the neovasculature quantitated by a fluorimetric method. Microvasc. Res. 54, 253–261.

Andrade, S.P., Hart, I., Piper, P.J., 1992. Inhibitors of nitric oxide synthase selectively reduce flow in tumour-associated neovasculature. Br. J. Pharmacol. 107, 1092–1095.

Belo, A.V., Barcelos, L.S., Teixeira, M.M., et al., 2004. Differential effects of antiangiogenic compounds in neovascularization, leukocyte recruitment, VEGF production, and tumor growth in mice. Cancer Invest. 22, 723–729.

Bollet, A.J., Goodwin, J.F., Simpson, W.F., et al., 1958. Mucopolysaccharide, protein and DNA concentration of granulation tissue induced by polyvinyl sponges. Proc. Soc. Exp. Biol. Med. 99, 418–421.

Buckley, A., Davidson, J.M., Kamerath, C.D., et al., 1985. Sustained release of epidermal growth factor accelerates wound repair. Proc. Natl. Acad. Sci. U.S.A. 82, 7340–7344.

Davidson, J.M., Klagsbrun, M., Hill, K.E., et al., 1985. Accelerated wound repair, cell proliferation and collagen accumulation are produced by a cartilage-derived growth factor. J. Cell. Biol. 199, 1219–1227.

Edwards, R.H., Sarmenta, S.S., Hans, G.M., 1960. Stimulation of granulation tissue growth by tissue extracts: study by intramuscular wounds in rabbits. Arch. Pathol. 69, 286–302.

Ferreira, M.A., Barcelo, L.S., Campos, P.P., et al., 2004. Sponge induced angiogenesis and inflammation in PAF receptor-deficient mice (PAFR-KO). Br. J. Pharmacol. 141, 1185–1192.

Grindlay, J.H., Waugh, J.M., 1951. Plastic sponge which acts as a framework for living tissue; experimental studies and preliminary report of use to reinforce abdominal aneurysms. AMA Arch. Surg. 63, 228–297.

Holund, B., Clemmensen, I., Junker, P., et al., 1982. Fibronectin in experimental granulation tissue. Acta Pathol. Immunol. Scand. (A) 99, 418–421.

Holund, B., Junker, P., Garbarsch, C., et al., 1979. Formation of granulation tissue in subcutaneously implanted sponges in rats. Acta Pathol. Microbiol. Scand. (A) 87, 367–374.

Kety, S.S., 1949. Measurement of regional circulation by local clearance of radioactive sodium. Am. Heart J. 38, 321–331.

Mahadevan, V., Hart, I.R., Lewis, G.P., 1989. Factors influencing blood supply in wound granuloma quantified by a new in vivo technique. Cancer Res. 49, 415–419.

Mc Grath, J.C., Arribas, S., Daly, C.J., 1996. Fluorescent ligands for the study of receptors. Trends Pharmacol. Sci. 17, 393–399.

Miyadera, K., Sumizawa, T., Haraguchi, M., et al., 1995. Role of thymidine phosphorylase in the angiogenic effect of platelet derived endothelial cell growth factor/thymidine phosphorylase. Cancer Res. 55, 1687–1690.

Paulini, K., Korner, B., Beneke, G., et al., 1974. A quantitative study of the growth of connective tissue: investigations on polyester–polyurethane sponges. Connect. Tiss. Res. 2, 257–264.

Pesenti, E., Sala, F., Mongelli, F., et al., 1992. Suramin prevents neovascularization and tumor growth through blocking of basic fibroblast growth factor activity. Br. J. Cancer 66, 367–372.

Thiede, K., Momburg, F., Zangemeister, U., et al., 1988. Growth and metastasis of human tumors in nude mice following tumor cell inoculation into a vascularized polyurethane sponge matrix. Int. J. Cancer 42, 939–945.

Wechsler, S., Fehr, D., Molenberg, A., et al., 2008. A novel, tissue occlusive poly(ethylene glycol) hydrogel material. J. Biomed. Mat. Res. Part A 85, 285–292.

Woessner Jr., J.F., Boucek, R.J., 1959. Enzyme activities of rat connective tissue obtained from subcutaneously implanted polyvinyl sponge. J. Biol. Chem. 234, 3296–3300.

CHAPTER 5

The Disk Angiogenesis System

THE EXPERIMENTAL MODEL

Fajardo et al. (1988 a,b) introduced the disk angiogenesis system (DAS) implanted subcutaneously in the host animal through a distal skin incision and then evaluated for penetration by host-derived blood vessels and/or other cell infiltrates (Fig. 5.1). The implant system consists of 11 mm disks of polyvinyl alcohol foam, covered on both sides by Millipore filters, leaving only the edge as the area for cell penetration in the disk. Angiogenic factors or antagonists, as well as other substances to be studied, are placed in the center of the disk. The slow release of these substances is assured by a film of ethylene−vinyl acetate copolymer, or by the use of agarose. In mice, the optimal times for examination of the disk are 7−12 days after implantation for disks containing proangiogenic factors, and 12−20 days for those without stimulators. After implantation, the disk is encapsulated by granulation tissue and invaded by new vessels.

The proliferation of endothelial and other cells is determined by incorporation of tritiated thymidine, using scintillation counting and autoradiography. Using the DAS, well-established angiogenic agents such a fibroblast growth factor-2 (FGF-2), epidermal growth factor (EGF) and prostaglandin E1 (PGE1) s were found to increase the proliferation of endothelial cells and microvessels. Heparin augmented the effect of FGF-2, while when used by itself heparin increased angiogenesis but not endothelial cell proliferation. Locally applied hyperthermia and ionizing radiation decreased angiogenesis, even when applied after the angiogenic stimulus (Fajardo et al., 1988a,b; Prionas et al., 1990). Systemic prostaglandin synthetase inhibitors antagonized the angiogenic effects of FGF-2 and EGF. Tumor necrosis factor alpha (TNF-α) tested at low concentrations induced angiogenesis, while at high concentrations inhibited angiogenesis (Fajardo et al., 1992). Transforming growth factor beta (TGF-β) at low concentrations had no effects, while at high concentrations induced significant angiogenesis (Fajardo et al., 1996). By using this assay, it has been established

In Vivo Models to Study Angiogenesis. DOI: http://dx.doi.org/10.1016/B978-0-12-814020-8.00005-6

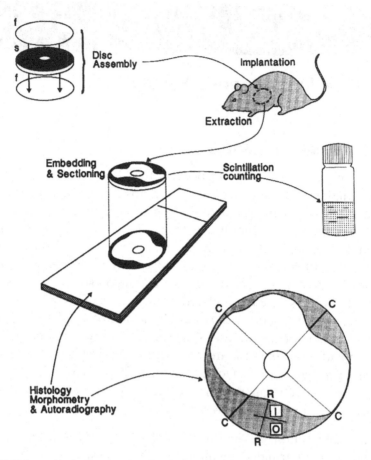

Figure 5.1 *Disk assembly, implantation removal, embedding sectioning and analysis.* Reproduced from Allison, A.C., Fajardo, L.F., 2011. Disc angiogenesis assay. Method. Mol. Med. 46, 59–75, 2011.

that glucocorticoids are effective in counteracting the angiogenic effects of FGF-2 and EGF (Ingber et al., 1986).

A similar approach was used by Bishop et al. (1990), who used subcutaneous implants of polyurethane foam cylinders. In these studies, cells were introduced into the cylinders prior to closing of the incision and 10 days after implantation, the cylinders were removed and processed for immunohistochemistry. The sponge of DAS can be coated before implantation with fibronectin, vibronectin, proteoglycan, various types of collagen, and other connective components.

The DAS has been modified to enable the introduction of live cells, i.e., tumor cells (glioma cells) or inflammatory cells into the center of disk (Nelson et al., 1993). The slow release of the test substance or of factors from the tumor cells is maintained by a film of ethylene–vinyl acetate copolymer or by the use of agarose (Fajardo et al., 1988a,b).

The disks are harvested after 1–3 weeks, fixed, sectioned, and stained. In the 14-day disks, there are two different growth zones: outer (peripheral) and inner (central) which can be separated by dividing the growth area into two equally wide bands. The outer zone usually contains a large amount of collagen continuous with that of the host tissue outside the disk, while the inner zone has a greater concentration of vessels.

Histological examination of sections of the disk shows a rich neovascularization and a distinctive cellular infiltration at the edges, including fibroblasts, endothelial cells, and leukocytes. Throughout the area of growth the majority of cells are fibroblasts, the second most common cells are endothelial cells. In the outer zone of the disk cellularity is high, while in the inner zone it is low (Fig. 5.2). New vessels

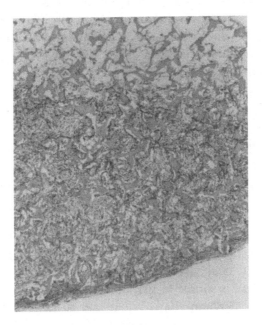

Figure 5.2 Section of DAS showing empty polyvinyl sponge in the upper one-third of the photograph (inner portion of disk). The lower two-thirds (outer or peripheral portion of disk) contain vessels and fibroblasts among the sponge trabeculae. The rim of the disk is at the bottom. Disk is stimulated by 20 pg of EGF, 14 days after implantation. Reproduced from Allison, A.C., Fajardo, L.F., 2011. Disc angiogenesis assay. Method. Mol. Med. 46, 59–75, 2011.

sprout from host venules with microvessels growing centripetally into the disk, together with fibroblasts. In the DAS there is always a moderate, well-characterized and measurable spontaneous vascular growth.

DAS does not contain in its fibrovascular growth any epithelial elements including glands, and all tubular structures lined by a basal lamina should be vessels. In this context, staining of collagen IV in the capillary basal lamina is the most precise method to outline the capillaries.

Electron microscopy can also be used alone or in combination with light microscopy. The visualization of vessels can be improved with intravascular dyes, inert substances (e.g., India ink), or microspheres. Moreover, radioactive isotopes can be employed to label cells or stroma and label within individual cell types can be detected by autoradiography.

Direct measurement of blood vessels can be performed with point counting on histological sections or determination of intravascular volume with radioactive isotopes (Mahadevan et al., 1989). Differential counting of the different component of the fibrovascular growth can be performed with microscope reticle, evaluating the absolute numbers of cells per given area (Kowalski et al., 1992). Morphometric analyses have shown that vessel growth is directly proportional to total fibrovascular growth.

Advantages
1. The capability of discriminating between proliferation and migration of endothelial cells and fibroblasts.
2. As the disk is an avascular structure at onset, all vessels are newly formed and the problem of differentiating new from preexisting vessels does not arise.
3. Multiple disks can be used for each time or dose point, which allows reproducible measurements of vascular growth and increases statistical accuracy.
4. The procedure is simple to prepare and inexpensive.
5. The DAS is a mammalian system and is therefore more relevant to human physiology and pathology.
6. Assesses wound healing and angiogenesis.
7. The DAS is easy to assemble and implant in small animals, including mice, which tolerate it well.

8. Since it is well tolerated, there is little inflammatory reaction that interferes with the angiogenesis induced by proangiogenic factors.
9. In the DAS assay vessels and stromal elements grow concomitantly. In this context, the assay can be used to examine the role of different stromal components in endothelial cell proliferation and migration.
10. The procedure is simple and inexpensive.

Disadvantages

1. The major drawback of this method is the inability to perform continuous in vivo monitoring of the angiogenic response. External and gross inspection cannot be performed. Each disk provides information for only one time point. Histological embedding, sectioning, and staining are required.
2. Encapsulated by granulation tissue.

REFERENCES

Allison, A.C., Fajardo, L.F., 2011. Disc angiogenesis assay. Method. Mol. Med. 46, 59–75.

Bishop, D.K., Sedmak, D.D., Leppink, D.M., et al., 1990. Vascular endothelial differentiation in sponge matrix allografts. Hum. Immunol. 28, 128–133.

Fajardo, L.F., Kwann, H.H., Kowalski, J., et al., 1992. Dual role of tumor necrosis factor-α in angiogenesis. Am. J. Pathol. 140, 539–549.

Fajardo, L.F., Prionas, S.D., Kwann, H.H., et al., 1996. Transforming growth factor β1 induces angiogenesis in vivo with a threshold pattern. Lab. Invest. 74, 600–608.

Fajardo, L.F., Kowalski, J., Kwan, H.H., et al., 1988a. The disc angiogenesis system. Lab. Invest. 58, 718–724.

Fajardo, L.F., Prionas, S.D., Kowalski, J., et al., 1988b. Hyperthermia inhibits angiogenesis. Radiat. Res. 114, 297–306.

Kowalski, J., Kwan, H.H., Prionas, S.D., et al., 1992. Characterization and applications of the disc angiogenic system. Exp. Mol. Pathol. 56, 1–19.

Ingber, D.E., Madri, J.A., Folkman, J., 1986. A possible mechanism for inhibition of angiogenesis by angiostatic steroids: induction of capillary basement membrane dissolution. Endocrinology 119, 1768–1775.

Mahadevan, V., Hart, I.R., Lewis, G.P., 1989. Factors influencing blood supply in wound granuloma quantitated by a new in vivo technique. Cancer Res. 49, 415–419.

Nelson, M.J., Conlet, P.K., Fajardo, L.F., 1993. Application of the disc angiogenesis system to tumor-induced neovascularization. Exp. Mol. Pathol. 58, 105–113.

Prionas, L.F., Kowalski, J., Fajardo, L.F., et al., 1990. Effects of x-irradiation on angiogenesis. Radiat. Res. 124, 43–49.

CHAPTER 6

The Dorsal Air Sac Model

THE EXPERIMENTAL MODEL

The dorsal air-sac model was developed by Selye (1953) as a means of monitoring the vascularization of tumor grafts. Air is injected under the dorsal skin of rats (or mice), the skin is lifted up from an area of white fascia, permitting the introduction of cells or tissue fragments temporarily creating a thin, isolated vascularized membrane for cells or tissues to establish a new blood supply (Fig. 6.1). In the dorsal air-sac model, simple implantation of a chamber ring loaded with tumor cells causes angiogenic vessel formation on the murine skin attached to the ring (Oikawa et al., 1997).

Both sides of a Millipore ring are covered by filters and the resultant chamber filled with a tumor cell suspension and then it is implanted into the preformed dorsal air sac of an anaesthetized mouse (Yonekura et al., 1999). The vascular response, which was apparent 48 h after tumor implantation, consisted of vasodilation and an increase in the number of visible vascular channels.

Following treatment with the compound of interest, the chamber is carefully removed and rings of the same diameters are placed directly upon the sites that were exposed to a direct contact with the chamber. The number of newly formed blood vessels that lie within the area marked by the ring is counted using a dissecting microscope. Care must be taken to not irritate the surface upon which the chamber is placed, as this may itself induce angiogenesis.

The rat air-sac model has been used to demonstrate in vivo the angiogenic activity of tumor angiogenesis factor (TAF) (Fig. 6.2) (Folkman et al., 1971) and in two studies by Cavallo et al. (1972, 1973). In the first one, Walker ascite tumor cells and TAF were injected into the fascial floor of the dorsal air sac. At intervals thereafter, ^{3}H-labeled thymidine was injected into the air sac and the tissues were examined by autoradiography and electron microscopy.

In Vivo Models to Study Angiogenesis. DOI: http://dx.doi.org/10.1016/B978-0-12-814020-8.00006-8

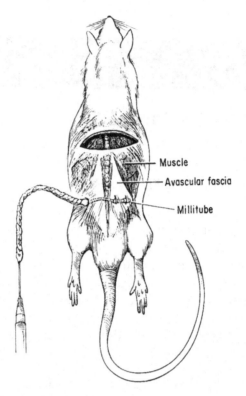

Muscle
Avascular fascia
Millitube

Figure 6.1 Schematic diagram of the rat dorsal air-sac assay. Reproduced from Folkman, J., Merler, E., Abernathy, C., et al., 1971. Isolation of a tumor factor responsible for angiogenesis. J. Exp. Med. 133, 275–288.

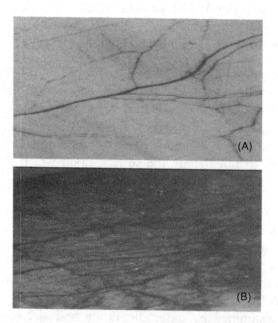

(A)

(B)

Figure 6.2 (A) Negative response in the rat dorsal air sac following 48 h exposure to millitube without TAF. (B) Strong angiogenic response following 48 h exposure to millitube with TAF. Reproduced from Folkman, J., Merler, E., Abernathy, C., et al., 1971. Isolation of a tumor factor responsible for angiogenesis. J. Exp. Med. 133, 275–288.

Autoradiographs showed thymidine labeling of endothelial cells, as early as 6–8 h after exposure to live tumor cells. DNA synthesis by endothelial cells subsequently increased, and within 48 h new vessels were detected. Further ultrastructural evidence showed that by 48 h there was ultrastructural evidence in endothelial cells, including marked increase in ribosomes and endoplasmic reticulum, scarce or absent pinocytotic vesicles, and discontinuous basement membrane (Cavallo et al., 1973).

Evans blue may be injected into the mice, which leaks out of the angiogenic vessels thereby accumulating in the interstitial spaces, but is retained within the preexisting vessels; the accumulation of dye is assessed by a semiquantitative measurement of angiogenesis (Yamakawa et al., 2004). It is possible to measure blood volume by determining the amount of ^{51}Cr-labeled erythrocytes circulating in the skin using a gamma counter (Funahashi et al., 1999).

Advantage
1. The procedure is simple.

Disadvantages
1. Nonspecific inflammatory response.
2. It is sometimes difficult to distinguish the preexisting vessels from the new vessels.
3. Inability to monitor the evolution of the angiogenic response.

REFERENCES

Cavallo, T., Sade, R., Folkman, J., et al., 1972. Tumor angiogenesis: rapid induction of endothelial mitosis demonstrated by autoradiography. J. Cell. Biol. 54, 408–420.

Cavallo, T., Sade, R., Folkman, J., et al., 1973. Ultrastructural autoradiographic studies of the early vasoproliferative response in tumor angiogenesis. Am. J. Pathol. 70, 345–362.

Folkman, J., Merler, E., Abernathy, C., et al., 1971. Isolation of a tumor factor responsible for angiogenesis. J. Exp. Med. 133, 275–288.

Funahashi, Y., Wakabayashi, T., Semba, T., et al., 1999. Establishment of a quantitative mouse dorsal air sac model and its application to evaluate a new angiogenesis inhibitor. Oncol. Res. 11, 319–329.

Oikawa, T., Sasaki, M., Inose, M., et al., 1997. Effects of cytogenin, a novel microbial product, on embryonic and tumor cell-induced angiogenic response in vivo. Anticancer Res. 17, 1881–1886.

Selye, H., 1953. On the mechanism through which hydrocortisone affects the resistance of tissues to injury; an experimental study with the granuloma pouch technique. J. Am. Med. Assoc. 152, 1207–1213.

Yamakawa, S., Asai, T., Uchida, T., et al., 2004. Epigallocatechin gallate inhibits membrane-type 1 matrix metalloproteinase, MT1-MMP, and tumor angiogenesis. Cancer Lett. 210, 47–55.

Yonekura, K., Basaki, Y., Chikahisa, L., et al., 1999. UTF and its metabolites inhibit the angiogenesis induced by murine renal cell carcinoma, as determined by a dorsal air sac assay in mice. Clin. Cancer Res. 5, 2185–2191.

CHAPTER 7

The Chamber Assays

THE EXPERIMENTAL MODEL

Since 1924 when the first rabbit-ear transparent chamber model was introduced by Sandison (1924), many other chamber models have been described for studying angiogenesis and microcirculation in a wide variety of neoplastic and nonneoplastic tissues by means of intravital microscopy, including the rabbit-ear chamber (Sandison, 1928), cranial chamber (Yuan et al., 1994), femur chamber (Hamsen-Algenstaedt et al., 2005), body windows (Bertera et al., 2003), and dorsal skinfold chamber (Algire, 1943).

The rabbit-ear chamber was adapted for use in mice to quantify structural and functional changes in the vasculature of tumors (Algire, 1943). In 1945, Algire and Chalkley used a transparent chamber implanted in a cat's skin to study the vasoproliferative reaction secondary to a wound or implantation of normal or neoplastic tissues. They showed that the vasoproliferative response induced by tumor tissues was more substantial and earlier than that induced by normal tissues or following a wound. They concluded that the growth of a tumor is closely connected to the development of an intrinsic vascular network.

Early experiments using in vivo microscopy on mammalian tissues were reviewed by Clark et al. (1931). A modification of the Sandison–Clark ear chamber which allowed the mica coverslip to be removed, exposing the surface of the table vascular membrane and then replaced, was adapted by Williams (1934, 1954) to study graft tissue fragments implanted on the membrane.

A piece of skin (ear and skinfold chambers) or part of the skull (cranial window chamber) is removed from an anaesthetized animal. In the dorsal skin chamber, one of the basic requirements is that the area of the dorsal skin associated with the chamber preparation lacks any injuries, scars, or other irritations. Before to chamber implantation, the back of the animal is shaved, a fold of depilated dorsal skin is

In Vivo Models to Study Angiogenesis. DOI: http://dx.doi.org/10.1016/B978-0-12-814020-8.00007-X

taken, and a circular area of one skin layer 15 mm thick is cut out. The skinfold is fixed between two titanium frames of the chamber and the operating field is covered with a sterile coverslip. Then, the coverslip is removed, cells and/or test agents inserted, and the chamber closed with a new sterile coverslip. Angiogenesis can be quantified in vivo by intravital microscopy or excised tissues can be studied by immunohistochemistry.

In the rabbit-ear chamber assay, the chamber is composed of a disk with a central round table, three peripheral pillars, a cover plate, and a holder ring (Fig. 7.1). Four holes are punched through the ear cartilage and skin with a special puncher. The four holes consist of three outer perforations and a central puncture that incorporate the transparent window. Three peripheral pillars of the disk are inserted through the outer small holes of the ear to position the chamber and the central round table plugs the large puncture in the center. A cover plate is fixed on the pillars by a holder ring. The disk is designed to leave a gap between the central round table and the cover plate. This space forms a cavity into which newly, sprouting microvessels can form the vessels of subdermis. After the operation, the chamber is filled with blood and cellular debris. New vessels originate by sprouting from pre-existing vessels that are adjacent to the cellular debris (Fig. 7.2).

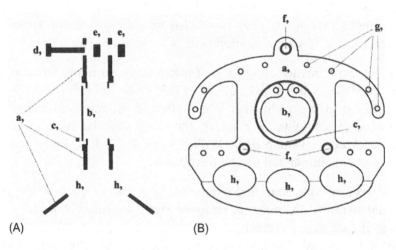

(A)   (B)

Figure 7.1 Construction plan of the dorsal skinfold chamber. (A) Cross section. (B) Lateral view. a, titanium frame; b, coverslip; c, tension ring; d, screw; e, nut; f, screw hole; g, bore holes for holding sutures; h, holes for weight reduction of the chamber. Reproduced from Sckell, A., Leunig, M., 2001. Dorsal skinfold chamber preparation in mice. Method. Mol. Med. 46, 95–105.

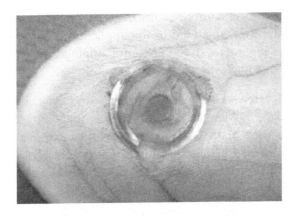

Figure 7.2 Rabbit-ear chamber. Arterioles and venules can be observed microscopically in the center window.

Tumor cells, or a gel containing angiogenic factors, are then placed on the exposed surface and covered by glass. The rabbit-ear chamber has been used to quantify structural and functional changes in the neovasculature of tumors (Durar and Jain, 1983). The dorsal skin chamber has been used to study tumor angiogenesis, including xenografts in immunodeficient rodents (Dellian et al., 1996). A filming system has been developed to obtain clear pictures of the chamber with good contrast, which allows to photograph neo-angiogenesis at the edge of the chamber by 1 week and fully developed vasculature by 4 weeks (Tomizawa et al., 2002).

Wood (1958) carried out detailed studies of tumor angiogenesis by using these chambers by means of a high-powered time-lapse film of rabbit-ear chambers. He showed that new capillary sprouts began to grow as early 18 h after a metastatic focus of tumor cells appeared in the extravascular space

The rabbit-ear chamber has been used to assess the angiostatic effects of corticosteroids on wound healing (Hashimoto et al., 2002) and microcirculatory changes during angiogenesis (Ichioka et al., 1997).

In an alternative method, two symmetrical frames are implanted into the dorsal skinfold to sandwich the extended double layer of the skin: one layer is completely removed exposing the other layer, which is protected by covering with a glass coverslip incorporated into one of the frames (Menger and Lahr, 1993). By using this method, it is possible to remove the coverslip for implantation of tumor cells or transplants into the camber ready for neovascularization (Laschke et al., 2006).

In 1980, the dorsal skinfold chamber work focused on microcirculation experimentation, which led to the applications for tumor research, especially with regard to studying tumor angiogenesis and antiangiogenic molecules (Fig. 7.3).

Transillumination techniques are used to study vessel diameter, density, and blood flow (Menger et al., 2002). Epi-illumination, fluorescence techniques are used to in vivo analysis of other parameters, including platelet adhesion, vascular permeability, and erythrocyte flow (Menger and Lahr, 1993; Bingle et al., 2006; Reyes-Aldasoro et al., 2008).

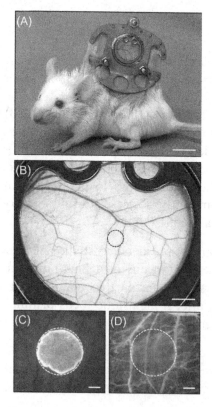

Figure 7.3 (A) BALB/c mouse with a dorsal skinfold chamber. (B) Observation window of a dorsal skinfold chamber directly after transplantation of a CT26 tumor cell spheroid (border marked by broken line). (C, D) Intravital fluorescence microscopy of the tumor cell spheroid (border marked by broken line) in (B). Because the cell nuclei of the spheroid were stained with the fluorescent dye Hoechst 33342 before transplantation, the implant can easily be differentiated from the nonstained surrounding host tissue of the chamber using ultraviolet light epi-illumination (C). Blue light epi-illumination of the identical region of interest as in (C) with contrast enhancement by intravascular staining of plasma with 5% FITC-labeled dextran 150,000 IV allows the visualization of the microvasculature surrounding the spheroid (D). Reproduced from Wittig, X., Scheuer, C., Parakenings, J., et al., 2015. Gerananial suppresses angiogenesis by downregulating vascular endothelial growth factor (VEGF)/ VEGFR-2 signaling. PLoS ONE 10, e131946.

In the cranial window preparation, a bone flap is prepared and the underlying dura is removed exposing the brain hemispheres (Fig. 7.4) (Yuan et al., 1994). Cranial window was originally developed for the direct observation of pial microcirculation (Levasseur et al., 1975). Tumor specimens, cell suspensions, or spheroids can be placed onto the pial surface of either hemisphere and the window sealed with a glass coverslip (Vajkoczy et al., 2002).

Other semitransparent preparations include the hamster cheek pouch (Klintworth, 1973) and the rat mesentery assay (Norrby et al., 1990). In Shubik's laboratory, the transparent chamber technique was adapted to the hamster cheek pouch to study angiogenesis in tumors (Goodall et al., 1965; Greenblatt and Shubik, 1968; Greenblatt et al., 1969; Shubik et al., 1976).

The hamster cheek pouch is considered in immunological privileged site where it is possible to grow allogenic or xenogenic grafts without inflammatory reaction, and it has been used to examine vascular reactions induced by tumor cells (Fig. 7.5). Observation of the transplant

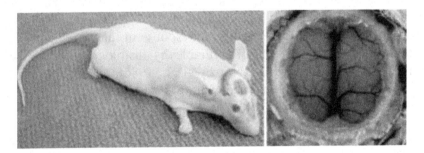

Figure 7.4 Mouse cranial window model. Nude mouse bearing a cranial window (left) and a macroscopic image of the mouse cranial window (right). The window is 7 mm across.

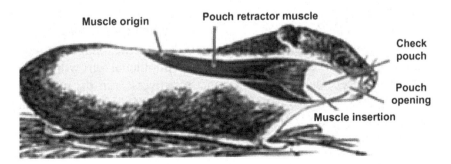

Figure 7.5 Scheme of the anatomical disposition of the hamster cheek pouch.

and changes in its vascular component is accomplished with an ordinary microscope equipped with long working distance objectives. The number of vessels in the region surrounding the implant is counted on histological sections.

Warren and Shubik (1966) used hamster cheek pouch to study the structure of tumor vessels of melanoma transplants. Greenblatt and Shubik (1968) implanted tumors in a Millipore chamber in the hamster cheek pouch and demonstrated that the tumor was capable of inducing new vessels on the opposite side of the Millipore filter. Since the pore size was 0.45 μm, cells could not traverse it and was assumed that some diffusible material crossed through the filter and induced neovascularization. Moreover, the hamster cheek pouch has been also used to determine the angiogenic potential of transforming growth factor alpha (TGF-α) (Schreiber et al., 1986).

The rat mesentery assay was originally described in 1986 (Norrby et al., 1986). The mesenteric window, which measures 5–10 μm in thickness, is covered on both sides by a single layer of mesothelial cells limiting a space in which are dispersed fibroblasts, mast cells, macrophages, and occasional eosinophils and lymphocytes (Norrby and Enestrom, 1984). The angiogenic response in the two-dimensional vasculature of the tissue is quantifiable by histological sections cut perpendicular to the surface (Norrby et al., 1986), and in spread preparations of intact, unsectioned windows where the entire vascular tree is visualized (Fig. 7.6) (Norrby et al., 1990).

The number of vessel profile per unit length of tissue is counted microscopically in stained sections, while in spread intact windows, the vascularized area (VA), as a percentage of the whole window area or in μm^2, and the vascular density (VD) within the VA are measured morphometrically, whereas the total vascular quantity is computed from VA per VD (Norrby et al., 1990). Norrby et al. (1986) induced angiogenesis by activating resident mast cells with the compound 48/80 (Fig. 7.7). Measurements of vascular endothelial growth factor (VEGF)–induced angiogenesis in intact microvessels in the mesentery of 5- to 6-week-old mice have been performed (Mukhopadhyay et al., 1998). Immunostaining of the microvascular network shows that the microvascular bed is composed of arterioles, capillaries, and venules.

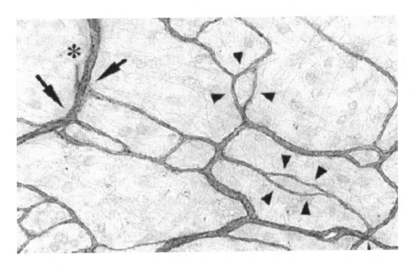

Figure 7.6 Following staining with a specific antibody antirat endothelium, the microvascular network of the rat mesentery is visualized. Asterisk, vascular sprout; arrows, interconnections; arrowheads, interconnecting loops. Reproduced from Norrby, K., 2006. In vivo models of angiogenesis. J. Cell. Mol. Med. 10, 588–612.

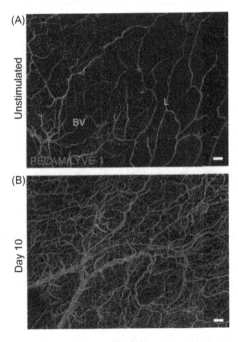

Figure 7.7 Mesenteric microvascular networks from (A) unstimulated and (B) compound 48/80-stimulated rats. Immunolabeling for Platelet endothelial cell adhesion molecule (PECAM-1) (red, dark gray in print versions) and Lymphatic vessel endothelial hyaluronan receptor 1 (LYVE1) (green, light gray in print versions) identified the hierarchies of branching blood vessel (BV) and lymphatic (L) networks. Evidence for angiogenesis and lymphangiogenesis by day 10 is supported by a dramatic increase in vessel density for both blood and lymphatic vessel networks. Reproduced from Sweat, R.S., Stapor, P.C., Murgee, W.L., 2012. Relationship between lymphangiogenesis and angiogenesis during inflammation in rat mesentery microvascular networks. Lymphatic. Res. Biol. 10, 198–207.

Dvorak et al. (1987) implanted fibrin gel contained in small perforated plastic chambers in the dorsal subcutaneous space of guinea pig. After 4 days of implantation, neovessels could be observed in the chamber containing the fibrin gel, which act as a chemotactic and mechanical support for neovascularization.

Advantages

1. Chamber assays allow for the determination of three-dimensional vessel growth in one animal, typically over a period of 1–3 weeks. Separate groups of mice are not required at each measurement point, and hence, the number of animals used is minimized.
2. The possibility of monitoring angiogenesis in vivo continuously up to several weeks, allowing for both real-time and time-lapse analysis. In addition, at the end of experiments, tissue samples can be excised and further examined by histology, immunohistochemistry, and molecular biology.
3. Transparent chamber enables determination of whether a newly formed blood vessel is perfused and contributed to tissue oxygenation.
4. Due to its extreme thinness, the intact mesenteric window is ideal for quantitative analyses of microvessel variables.

Disadvantages

1. All chamber assays are invasive and technically demanding. The surgical procedure can induce a secondary angiogenic response through wound healing that could be superimposed on the effects of the test substance themselves. Moreover, the presence of newly formed granulation tissue within the chamber influences the results obtained with the tumor implants.
2. The skinfold chamber is not an orthotopic site for many of the tumor studied. Therefore, the cranial window preparation was generated to provide an orthotopic brain tumor model (Yuan et al., 1994).
3. The skinfold chamber can have poor optical properties due to skin thickness and the cranial window requires the injection of a florescent marker in order to visualize new vessels.
4. Expensive.
5. Nonspecific inflammatory response.
6. The skinfold chamber can be used for 4–6 weeks after which loss of skin elasticity and necrosis at sutured sites affects chamber viability.
7. In the rabbit-ear chamber assay, the surgical procedure is technically demanding.

REFERENCES

Algire, G.H., 1943. An adaptation of the transparent chamber technique to the mouse. J. Natl. Cancer. Inst. 4, 1–11.

Algire, G.H., Chalkley, H.W., 1945. Vascular reactions of normal and malignant tissue in vivo. J. Natl. Cancer Inst. 6, 73–85.

Bertera, S., Geng, X., Tawadrous, Z., et al., 2003. Body window-enabled in vivo multicolor imaging of transplanted mouse islet expressing an insulin-timer fusion protein. Bio-Techniques 35, 718–722.

Bingle, L., Lewis, C.E., Corke, K., et al., 2006. Macrophages potentiate angiogenesis in the dorsal skinfold chamber model. Br. J. Cancer 94, 101–107.

Clark, E.R., Hitschler, W.J., Kirby Smith, H.T., et al., 1931. General observations on ingrowth of new blood vessels into the standardized chambers in rabbit's ear, and subsequent changes in newly grown vessels over period of months. Anat. Rec. 50, 129–151.

Dellian, M., Witwer, B.P., Salehi, H.A., et al., 1996. Quantitation and physiological characterization of angiogenic vessels in mice: effect of basic fibroblast growth factor, vascular endothelial growth factor/vascular permeability factor, and host microenvironment. Am. J. Pathol. 149, 59–72.

Durar, T.E., Jain, R.K., 1983. Microcirculation flow changes during tissue growth. Microvasc. Res. 25, 1–21.

Dvorak, H.F., Harvet, V.S., Estrella, P., et al., 1987. Fibrin containing gels induce angiogenesis. Implications for tumor stroma generation and wound healing. Lab. Invest. 57, 673–686.

Goodall, C.M., Sanders, A.G., Shubik, P., 1965. Studies of vascular patterns in living tumors with a transparent chamber inserted in hamster cheek pouch. J. Natl. Cancer Inst. 35, 497–521.

Greenblatt, M., Choudari, K.V.R., Sanders, A.G., et al., 1969. Mammalian microcirculation in the living animals: methodological considerations. Microvasc. Res. 1, 420–432.

Greenblatt, M., Shubik, P., 1968. Tumor angiogenesis: transfilter diffusion studied in the hamster by the transparent chamber technique. J. Natl. Cancer Inst. 41, 111–124.

Hamsen-Algenstaedt, N., Schaefer, C., Wolfram, L., et al., 2005. Femur window, a new approach to microcirculation of living bone in situ. J. Orthop. Res. 23, 1073–1082.

Hashimoto, I., Nakanishi, H., Shono, Y., et al., 2002. Angiostatic effects of corticosteroids on wound healing of the rabbit ear. J. Med. Invest. 49, 61–66.

Ichioka, S., Shibata, M., Kosaki, K., et al., 1997. Effects of shear stress on wound-healing angiogenesis in a rabbit ear chamber. J. Surg. Res. 72, 29–35.

Klintworth, G.K., 1973. The hamster cheek pouch: an experimental model of corneal neovascularization. Am. J. Pathol. 73, 691–710.

Laschke, M.W., Elitzsch, A., Vollmar, B., et al., 2006. Combined inhibition of vascular endothelial growth factor (VEGF), fibroblast growth factor, but not inhibition of VEGF alone, effectively suppresses angiogenesis and vessel maturation in endometriotic lesions. Hum. Reprod. 21, 262–268.

Levasseur, J.E., Wei, E.P., Raper, A.J., et al., 1975. Detailed description of a cranial window technique for acute and chronic experiments. Stroke 6, 308–317.

Menger, M.D., Lahr, H.A., 1993. Scope and perspectives of intravital microscopy-bridge over from in vitro to in vivo. Immunol. Today 14, 519–522.

Menger, M.D., Laschke, M.W., Vollmar, B., 2002. Viewing the microcirculation through the window: some twenty years experience with the hamster dorsal skinfold chamber. Eur. Surg. Res. 34, 83–91.

Mukhopadhyah, D., Nagy, J.A., Manseau, E.J., et al., 1998. Vascular permeability factor/vascular endothelial growth factor-mediated signaling in mouse mesentery vascular endothelium. Cancer Res. 58, 1278–1284.

Norrby, K., Enestrom, S., 1984. Cellular and extracellular changes following mast cell secretion in avascular rat mesentery. An electron microscopic study. Cell. Tissue Res. 235, 339–345.

Norrby, K., 2006. In vivo models of angiogenesis. J. Cell. Mol. Med. 10, 588–612.

Norrby, K., Jakobsson, A., Sörbo, J., 1986. Mast cell-mediated angiogenesis: a novel experimental model using the rat mesentery. Virchow Arch. [Cell. Pathol.] 52, 195–206.

Norrby, K., Jakobsson, A., Sörbo, J., 1990. Quantitative angiogenesis in spreads of intact rat mesenteric windows. Microvasc. Res. 39, 341–348.

Reyes-Aldasoro, C.C., Akerman, S., Tozer, G.M., 2008. Measuring the velocity of fluorescent labeled red blood cells with a keyhole tracking algorithm. J. Microsc. 229, 162–173.

Sandison, J.C., 1924. A new method for the microscopic study of living growing tissues by introduction of a transparent chamber in the rabbit's ear. Anat. Rec. 28, 281–287.

Sandison, J.C., 1928. Transparent chamber of rabbit's ear, giving complete description of improved technic of construction and introduction, and general account of growth and behavior of living cells and tissues as seen with the microscope. Am. J. Anat. 41, 447.

Schreiber, A.B., Winkler, M.E., Derynck, R., 1986. Transforming growth factor alpha: a more potent angiogenic factor than epidermal growth factor. Science 232, 1250–1253.

Sckell, A., Leunig, M., 2001. Dorsal skinfold chamber preparation in mice. Method. Mol. Med. 46, 95–105.

Shubik, P., FeldmanR, Garcia, H., et al., 1976. Vascularization induces in the cheek pouch of the Syrian hamster by tumor and non tumor substances. J. Natl. Cancer Ints. 57, 69–774.

Sweat, R.S., Stapor, P.C., Murgee, W.L., 2012. Relationship between lymphangiogenesis and angiogenesis during inflammation in rat mesentery microvascular networks. Lymphatic. Res. Biol. 10, 198–207.

Tomizawa, Y., Suzuki, Y., Miyama, A., et al., 2002. Microscopic sequential pictures of angiogenesis in a rabbit ear chamber. J. Invest. Surg. 15, 269–274.

Vajkoczy, P., Farhadi, M., Gaumann, A., et al., 2002. Microtumor growth initiates angiogenic sprouting, with simultaneous expression of VEGF, VEGF receptor-2, and angiopoietin-2. J. Clin. Invest. 109, 77–785.

Warren, B.A., Shubik, P., 1966. The growth of the blood supply to melanoma transplants in the hamster cheek pouch. Lab. Invest. 15, 464–478.

Williams, R.G., 1934. Transparent chamber adapted for cell culture and permitting access to contained living tissue. Anat. Rec. 60, 487.

Williams, R.G., 1954. Microscopic studies in living mammals with transparent chamber methods. Int. Rev. Cytol. 3, 359–398.

Wittig, X., Scheuer, C., Parakenings, J., et al., 2015. Gerananial suppresses angiogenesis by downregulating vascular endothelial growth factor (VEGF)/VEGFR-2 signaling. PLoS One 10, e131946.

Wood, S., 1958. Pathogenesis of metastasis formation in vivo in the rabbit ear chamber. Arch. Pathol. 66, 550–568.

Yuan, F., Salehi, H.A., Boucher, Y., et al., 1994. Vascular permeability and microcirculation of gliomas and mammary carcinomas transplanted in rat and mouse cranial windows. Cancer Res. 54, 4564–4568.

CHAPTER 8

The Zebrafish

THE EXPERIMENTAL MODEL

The teleost zebrafish (*Danio rerio*) is a tropical freshwater species named for the five horizontal pigmented stripes that run longitudinally across their bodies, approximately 3–4 cm long as adult, with a short generation time of about 3 months that can be hosed in large numbers (Fig. 8.1). The zebrafish possesses a complex circulatory system (Fig. 8.2) similar to that of mammals (Weinstein, 2002), and the basic vascular plan of the developing zebrafish embryo shows strong similarity to that of other vertebrates (Isogai et al., 2001). At the 13 somite-stage, endothelial cell precursors migrating from the lateral mesoderm originate the zebrafish vasculature and a single blood circulatory loop is present at 24 h postfertilization (hpf). The dorsal aorta and posterior cardinal veins are formed by vasculogenesis. Blood vessel development continues during the subsequent days by angiogenic processes. Angiogenesis occurs in the formation of the intersegmental vessels (ISVs) of the trunk that will sprout from the dorsal aorta at 20 hpf (Zhang et al., 2001). Also, subintestinal vein vessels (SIVs) originate from the duct of Cuvier area at 48 hpf and will form a vascular plexus across most of the dorsal–lateral aspect of the yolk ball during the next 24 h (Isogai et al., 2001).

The vessel architecture of zebrafish embryos consists of a dorsal aorta and a posterior cardinal vein on each side. A network of secondary branches originates from these vessels, including dorsoventrally aligned ISVs, two separate longitudinal anastomotic vessels joining the ISVs and longitudinal parachordal vessels (Eriksson and Löfberg, 2000). Development of ISVs occurs at predicted time points and these vessels appear in an extremely well organized and regular pattern. The vasculature is composed of a single cell layer of endothelial cells, with smooth muscle cells, pericytes, and fibroblasts (Ellert et al., 2010).

Previous studies had shown that developmental angiogenesis in the zebrafish embryo, leading to the formation of the ISVs of the trunk

In Vivo Models to Study Angiogenesis. DOI: http://dx.doi.org/10.1016/B978-0-12-814020-8.00008-1

Figure 8.1 Zebrafish (D. rerio).

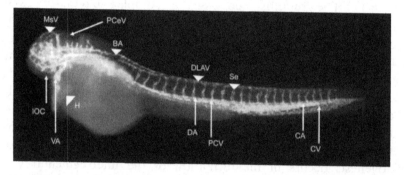

Figure 8.2 Vasculature of Tg(flk1:EGFP) zebrafish at 48 hpf. BA, Basilar artery; CA, caudal artery; CV, caudal vein; DA, dorsal aorta; DLAV, dorsal longitudinal anastomotic vessel; H, heart; IOC, inner optic circle; MsV, mesencephalic vein; PCeV, posterior cerebral vein; PCV, posterior cardinal vein; Se, intersegmental vessel. Reproduced from Lee, S.L., 2011. The Zebrafish as a Model to Elucidate Human Diseases and Development. Karolinska Institutet Stockholm.

(Cross et al., 2003) and of the SIV plexus (Serbedzija et al., 2000), represents a target for the screening of antiangiogenic compounds. In these assays, low-molecular-weight compounds dissolved in fish water are investigated for their impact on the growth of new blood vessels driven by the complex network of endogenous, developmentally regulated signals. Changes in the vasculature can be quantified by assessing the number of ISV and by measuring the extent of their outgrowth, while antiangiogenic agents may reduce the number or outgrowth of these vessels.

Zebrafish angiogenesis assays are conducted by injecting a test substance (protein/peptide) into the yolk sac of the embryos, while lipophilic test substances added to the water can freely diffuse into the embryos. The transparency allows to measure angiogenesis by visual inspection (Isogai et al., 2001). Small molecules inhibit angiogenesis already after only 24 h of development, reducing the assay time compared to other angiogenesis models. Chan et al. (2002) inhibited vascular endothelial growth factor receptor (VEGFR) signaling with the kinase inhibitor PTK787 and found a dose-dependent reduction in vascularization in the zebrafish embryos, which is counteracted by the overexpression of the VEGF-targeted PKB/Akt. The zebrafish assay could differentiate antiangiogenic compounds acting by VEGFR inhibition and those acting by nonselective cytotoxicity, like fumagillin (Chimote et al., 2014).

While common antisense technologies were not generally applicable to the zebrafish, the advent of oligonucleotide substitutes named morpholinos (MO) enabled the knockdown of endogenous genes by either blocking translation of the mRNA or splicing of the pre-mRNA. MO are synthetic RNA analogs that are stable in the cytosol and specifically bind to the designed target mRNA transcripts and thus blocking their translation.

MO "knockdown" technology for reverse genetic analysis of gene functions has been utilized as an useful approach to understanding molecular events in vascularization in the zebrafish model. Utilization of mutagens and antisense MO oligonucleotides facilitates genetic engineering for investigating the molecular mechanism of angiogenesis (Currie and Ingham, 1996). In this context, it has been generated mosaic fish embryos where cells from a genetically changed fish are isolated and injected into blastula stage wild type or differently genetically altered fish strains (Carmany-Rampey and Moens, 2006).

As the embryos develop outside the mother and are transparent, direct observation and quantification of blood vessel formation may be performed using a low-power microscope (Lawson and Weinstein, 2002a,b). Patent vasculature is visualized through injection of fluorescent dye, quantum dots, or microspheres (Leong et al., 2010; Lewis et al., 2006), followed by confocal microscopy and image reconstruction. Transgenic zebrafish with green fluorescent vasculature allows an easy visualization of the vascular tree (Weinstein et al., 1995). Of these different transgenics, the Fli-eGFP has the strongest fluorescence (Lawson and Weinstein, 2002a,b). A combination of the Fli-eGFO zebrafish injected with red fluorescent microspheres enables the observation of flow within the newly forming vessels (Chico et al., 2008).

A novel zebrafish yolk membrane (ZFYM) assay has been proposed based on the injection of an angiogenic growth factor, e.g., recombinant fibroblast growth factor-2 (FGF-2), in the perivitelline space of zebrafish embryos in the proximity of developing SIVs. FGF-2 induces a rapid and dose-dependent angiogenic response from the SIV basket, characterized by the growth of newly formed, alkaline phosphatase—positive blood vessels (Nicoli et al., 2008a). The ZFYM assay differs from the previous zebrafish-based angiogenesis assays since the angiogenic stimulus is represented by a well-defined, topically delivered exogenous agent that leads to the growth of ectopic blood vessels, allowing the screening of low- and high-molecular-weight antagonists targeting a specific angiogenic growth factor and/or its receptor(s) (Nicoli et al., 2008a).

ANGIOGENESIS IN ADULT ZEBRAFISH

Cao et al. (2008) developed a hypoxia-induced retinal angiogenesis model in the adult zebrafish to study pathological angiogenesis. They demonstrated that hypoxia can induce neovascularization in the retina of adult zebrafish, and that such neovascularization is dependent on the VEGF signaling. Moreover, orally active antiangiogenic drugs block the hypoxia-induced angiogenesis. In another experimental approach, partial fin amputation could serve as a model for regenerative angiogenesis (White et al., 1994). The availability of genetically modified zebrafish with green fluorescent endothelial cells made this model attractive for vital imaging of the postinjury angiogenesis. This regeneration assay has been introduced as nonembryonic angiogenesis model (Bayliss et al., 2006) and improves the discovery of genes and drugs related to angiogenesis (De et al., 2006).

ANGIOGENESIS IN EXPERIMENTAL ZEBRAFISH TUMORS

Zebrafish spontaneously develops almost any type of tumor, most commonly in testis, gut, thyroid, liver, peripheral nerve, connective tissue, and ultimobranchial gland. Also, several approaches have been developed to induce cancer in zebrafish. They include treatment with chemical carcinogens, forward genetic screening, target-selected inactivation of tumor suppressor genes, and expression of mammalian oncogenes (Feitsma and Cuppen, 2008; Stoletov and Klemke, 2008). Notably, microarray analysis has shown that gene expression signatures are conserved in fish tumors when compared to their human counterpart (Lam et al., 2006). Ultrasound biomicroscopy has been used to follow the growth of liver tumors, their vascularity, and response to treatment (Goessling et al., 2007). Other imaging techniques, including microcomputerized axial tomography, micromagnetic resonance imaging, and optical projection tomography have been applied in zebrafish and will help to investigate tumor growth and vascularization (Spitsbergen, 2007).

Relevant to tumor angiogenesis studies in zebrafish adults, a transparent *casper* zebrafish line that lacks all types of pigments has been generated (White et al., 2008). Crossing of the *casper* mutant with transgenic lines that label vasculature or internal organs with fluorescent tags may represent an useful approach to study tumor–host interactions in zebrafish by epifluorescence stereomicroscopy, confocal microscopy, and dual-photon confocal microscopy.

TUMOR ENGRAFTING AND ANGIOGENESIS IN ZEBRAFISH EMBRYOS

Because of the immaturity of the immune system in zebrafish embryos, no xenograft rejection occurs at this stage (Taylor and Zon, 2009; Weinstein, 2002). Moreover, transient gene inactivation via MO injection represents a powerful tool for the identification of target genes in zebrafish embryo (Thisse and Zon, 2002).

Original studies have shown the feasibility of injecting human melanoma cells in zebrafish embryos at blastula stage to follow their fate and to study their impact on zebrafish development (Topczewska et al., 2006). Grafting of tumor cells at this stage, well before vascular development, results in their reprogramming toward a nontumorigenic phenotype, thus hampering any attempt to investigate tumor-driven vascularization. At variance, injection of melanoma cells into the hindbrain ventricle or yolk sac of 48 hpf embryos results in the formation of tumor masses within 4 days (Haldi et al., 2006). Immunostaining analysis of the grafts reveals the presence of blood vessels within the brain and abdominal lesions, even though the high vascularity of the invaded regions may not allow easy discrimination between developmental and tumor-induced angiogenesis (Haldi et al., 2006).

An easier assessment of the angiogenic response triggered by tumor cell engrafting has been obtained by injecting mammalian tumor cells in the proximity of the developing SIV plexus in zebrafish embryos at 48 hpf (Nicoli and Presta, 2007; Nicoli et al., 2007). Proangiogenic factors released locally by the tumor graft, including FGF-2 and VEGF, affect the normal developmental pattern of the SIVs by stimulating the migration and growth of sprouting vessels toward the implant. One to two days after tumor cell grafting, whole-mount phosphatase alkaline staining allows the macroscopic evaluation of the angiogenic response (Fig. 8.3).

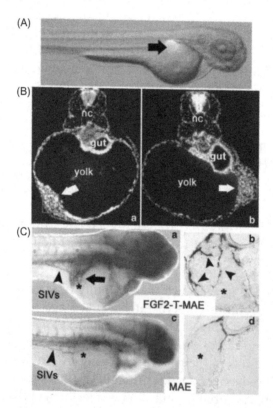

Figure 8.3 Tumor cell xenografts induce angiogenesis in zebrafish embryo. (A) Zebrafish embryo grafted with GFP-transduced cells to highlight the site of injection (arrow). (B) Tumorigenic murine FGF-2-T-MAE cells (a) and parental MAE cells (b) were injected into the perivitelline space of zebrafish embryos at 48 hpf. After 24 h, transverse sections of the embryos were stained with DAPI to visualize the cell graft (arrow). (C) Alkaline phosphatase staining of FGF-2-T-MAE cell–injected (a and b) and MAE cell–injected (c and d) embryos at 72 hpf. (a and c) Whole-mount lateral view showing numerous neovessels originating from the SIV basket that migrate and infiltrate the FGF-2-T-MAE cell graft (arrow in a), whereas no neovessel formation is observed in embryos injected with MAE cells (c). (b, d) Transverse sections of the cell grafts; alkaline phosphatase–positive intratumor blood vessels in the FGF-2-T-MAE cell graft. Reproduced from Nicoli, S., Ribatti, D., Cotelli, F., et al., 2007. Mammalian tumor xenografts induce neovascularization in zebrafish embryos. Cancer Res. 67, 2927–2931.

The use of transgenic zebrafish embryos, in which endothelial cells express green fluorescent protein (GFP) under the control of endothelial-specific promoters (Baldessari and Mione, 2008), represents an improvement of the zebrafish embryo/tumor xenograft model, allowing the observation and time-lapse recording of newly formed blood vessels in live embryos by epifluorescence microscopy as well as by in vivo confocal microscopy (Nicoli and Presta, 2007; Nicoli et al., 2007). Also, quantum dots may be used as labeling agents of the zebrafish embryo vasculature for long-lasting intravital time-lapse studies (Rieger et al., 2005).

When cancer cells are injected in zebrafish embryos maintained in hypoxic water, invasion into neighboring tissues, dissemination, and metastasis of labeled tumor cells was greatly enhanced when compared to cells injected under normoxic conditions (Lee et al., 2009). Accordingly, VEGFR blockade by sunitinib administration in the fish water or by anti-VEGFR-2 MO injection inhibited hypoxia-mediated pathological angiogenesis, early dissemination of malignant cells, invasiveness, and metastasis. The possibility to study the metastatic behavior of primary human tumors in zebrafish embryos has been confirmed also when human tumor tissue samples or primary tumor cells are injected into the yolk of 2-day-old embryos or are implanted into the liver of zebrafish larvae (Marques et al., 2009).

In zebrafish embryos, MO injection induces a translational block in gene function (Nasevicius and Ekker, 2000). Gene inactivation by this approach is easy and fast (3–4 days) when compared to the generation of knockout mice (several months). Also, the simultaneous injection of different MOs may allow the inactivation of more than one gene at the same time. This represents an advantage compared to any mammalian assay available and it can be exploited for the identification of novel gene(s) involved in tumor neovascularization. For instance, MO-induced inactivation of vascular endothelial (VE)-cadherin (Fig. 8.4) (Nicoli et al., 2007) or calcitonin receptor-like receptor (Nicoli et al., 2008b) zebrafish gene orthologs result in a significant inhibition of the angiogenic process triggered by the tumor graft in zebrafish embryos. On the other hand, silencing of LIM kinases in human pancreatic cancer cells by siRNA treatment results in a decreased angiogenic potential of these cells when tested in the zebrafish xenograft angiogenesis assay (Vlecken and Bagowski, 2009).

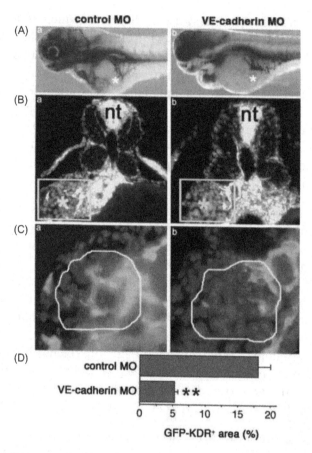

Figure 8.4 VE-cadherin MO inhibits tumor angiogenesis in zebrafish embryo. (A) Whole-mount alkaline phosphatase staining of FGF-2-T-MAE cell–grafted VEGFR2:G-RCFP control (a) and VE-cadherin (b) morphants at 72 hpf (lateral view). (B) DAPI-stained transverse sections of tumor-injected control (a) and VE-cadherin (b) morphants (nt, neuronal tube; *, cell graft; yellow (light gray in print versions) boxes, areas enlarged in (C)). (C) GFP-KDR⁺ neovessels (green, dark gray in print versions) within the tumor graft (white line) in VEGFR2: G-RCFP control (a) and VE-cadherin (b) morphants. (D) Quantification by computerized image analysis of the GFP-KDR⁺ area in tumors of control and VE-cadherin morphants. Reproduced from Nicoli, S., Ribatti, D., Cotelli, F., et al., 2007. Mammalian tumor xenografts induce neovascularization in zebrafish embryos. Cancer Res. 67, 2927–2931.

Because of the permeability of its embryos to small molecules, zebrafish allows disease-driven drug target identification and in vivo validation, thus representing an interesting bioassay tool for small molecule testing and dissection of biological pathways alternative to other vertebrate models (Pichler et al., 2003). Systemic exposure of live zebrafish embryos to antiangiogenic compounds dissolved in fish water results in a significant inhibition of neovascularization triggered by the tumor graft (Nicoli et al., 2007; Serbedzija et al., 1999). Also, nanoparticle-mediated targeting of intracellular signaling pathways (Harfouche et al., 2009) and novel anticancer metallodrugs (Brittijn et al., 2009) have been demonstrated to inhibit angiogenesis in this model.

The response of this model to angiogenesis inhibitors (24–48 h) is rapid when compared to other in vivo assays (Hasan et al., 2004). A large number of zebrafish embryos can be injected and maintained in 96 well plates, thus allowing systemic in vivo treatment of the animals with minimal amounts of compound. Therefore, dose–response experiments can be easily performed and numerous compounds can be tested.

The metabolic fate of the drug may differ in zebrafish embryo with respect to mammalian species. Zebrafish embryos are maintained at 28°C. This may not represent an optimal temperature for mammalian cell growth and metabolism, even though mitotic figures with no sign of apoptosis in grafted tumors throughout the whole experimental period have been observed (Nicoli et al., 2007). In this respect, the possibility to raise the incubation temperature up to 35°C with no apparent gross effects on zebrafish development has been reported (Haldi et al., 2006).

TUMOR ENGRAFTING AND ANGIOGENESIS IN ADULT AND JUVENILE ZEBRAFISH

Tumor transplantation in zebrafish has developed as an important tumor model (Taylor and Zon, 2009), a major limitation of this approach being the rejection of the tumor graft by the immune system of the host. Transplantable tumor cell lines have been generated in zebrafish and maintained for several passages in syngeneic and isogeneic adults (Mizgirev and Reskoy, 2006; Mizgirev and Revskoy, 2010). This should allow the study of tumor–endothelial cell interactions in immunocompetent zebrafish adults and juveniles. As an alternative approach, immunosuppression by dexamethasone administration (Stoletov et al., 2007) or sublethal gamma irradiation (White et al., 2008) can be used to prevent the rejection of tumor xenografts (Taylor and Zon, 2009). Also in this case, the use of transparent *casper* zebrafish may allow the rapid identification of transplanted tumor cells. Indeed, intraperitoneum or intraventricular injection of small numbers of GFP-labeled, transgenic zebrafish melanoma cells in irradiated *casper* recipients has allowed the study of three-dimensional tumor growth and whole body distribution of tumor cells, thus providing a quantitative analysis of tumor engraftment and offering the potential for monitoring in vivo effects of therapeutically useful molecules (White et al., 2008).

Human cancer cells have been successfully transplanted also in the peritoneal cavity of 30-day-old zebrafish juveniles (Stoletov et al., 2007), allowing the study of the dynamics of microtumor formation and neovascularization using high-resolution imaging techniques, leading to a detailed description of the interaction among fluorescent tumor cells and the GFP-labeled vasculature of the host by three-dimensional reconstruction of confocal microscopy images. This model system provides a clear window to visualize mechanisms of microtumor formation, cell invasion, and tumor-induced angiogenesis in a mature animal. However, due to the fact that juvenile zebrafish has a functional immune system, dexamethasone administration was required to prevent the rejection of the tumor engraftment. Also, at variance with zebrafish embryos, the MO gene targeting approach is unfeasible in zebrafish juveniles. On the other hand, the impact of the tumor graft on the mature vasculature of juvenile fishes may recapitulate more closely the events that occur during tumor angiogenesis in adult animals and cancer patients. Indeed, developing vessels of zebrafish embryos may respond differently to tumor grafts compared to the fully developed vasculature of juvenile animals (Stoletov and Klemke, 2008).

Advantages

1. The optical transparency allows direct and continuous microscopic inspection. The embryos remain transparent for a relatively long term, and full pigmentation is not achieved after several weeks. Moreover, the early development of a cardiovascular system in the transparent embryo translates into a unique opportunity for direct observation of blood flow and the development of the system's related organs in both wild-type and transgenic fish, without the need for complex instrumentation.
2. The model allows the delivery of a very limited number of cells, mimicking the initial stages of tumor angiogenesis and metastasis.
3. Labeled tumor cells (e.g., GFP-transduced or fluorescent dye-loaded cells) can be easily visualized within the embryo, allowing

analysis of the spatial/temporal relationship among tumor cells and newly formed blood vessels (Lawson and Weinstein, 2002a,b).

4. Several techniques can be applied including histochemistry, immunohistochemistry, and electron microscopy. Moreover, reverse transcriptase polymerase chain reaction analysis with species-specific primers allows the study of gene expression by grafted tumor cells and by the host under different experimental conditions (Nicoli et al., 2007).

5. Zebrafish cancer assays can be scaled up into medium- and high-throughput screens (Brittijn et al., 2009; Snaar-Jagalska, 2009).

6. The zebrafish share many genes and mechanisms of angiogenesis regulation with mammals (Rubinstein, 2003). Genetic studies have revealed conservation of the molecular pathways between fish and mammals making research in vascular biology in teleosts directly translatable into potentially relevant information for human health.

7. The embryos are inexpensive to generate and easy to maintain long term. Low costs are required to manipulate fish embryos and maintain the facilities as compared to rodents.

8. The experiments are relatively short, requiring only small amounts of drug per experiment.

9. The possibility to set up large-scale screening of small molecules makes the zebrafish model useful for the discovery and in vivo validation of new antiangiogenic molecules, as demonstrated by successful testing of known antiangiogenic compounds (Cao et al., 2008; van Rooijden et al., 2010).

Disadvantage
1. Nonmammalian.

REFERENCES

Baldessari, D., Mione, M., 2008. How to create the vascular tree? (Latest) help from the zebrafish. Pharmacol. Ther. 118, 206–230.

Bayliss, P.E., Bellavance, K.L., Whitehead, G.G., et al., 2006. Chemical modulation of receptor signaling inhibits regenerative angiogenesis in adult zebrafish. Nat. Chem. Biol. 2, 265–273.

Brittijn, S.A., Duivesteijn, S.J., Belmamoune, M., et al., 2009. Zebrafish development and regeneration: new tools for biomedical research. Int. J. Dev. Biol. 53, 835–850.

Cao, R., Jensen, L.D., Söll, I., et al., 2008. Hypoxia-induced retinal angiogenesis in zebrafish as a model to study retinopathy. PLoS One 3, e2748.

Carmany-Rampey, A., Moens, C.B., 2006. Modern mosaic analysis in the zebrafish. Methods 39, 228–238.

Chan, J., Bayliss, P.E., Wood, J.M., et al., 2002. Dissection of angiogenic signaling in zebrafish using a chemical genetic approach. Cancer Cell 1, 257–267.

Chico, T.J., Ingham, P.W., Crossman, D.C., 2008. Modeling cardiovascular disease in the zebrafish. Trend. Cardiovasc. Med. 18, 150–155.

Chimote, G., Sreenivasan, J., Pawar, N., et al., 2014. Comparison of effects of anti-angiogenic agents in the zebrafish efficacy-toxicity model for translational anti-angiogenic drug discovery. Drug. Des. Dev. Ther. 8, 1107–1123.

Cross, L.M., Cook, M.A., Lin, S., et al., 2003. Rapid analysis of angiogenesis drugs in a live fluorescent zebrafish assay. Arterioscler. Thromb. Vasc. Biol. 23, 911–912.

Currie, P.D., Ingham, P.W., 1996. Induction of specific muscle cell type by a hedgehog-like protein in zebrafish. Nature 382, 452–455.

De, S.F., Carmeliet, P., Autiero, M., 2006. Fishing and frogging for anti-angiogenic drugs. Nat. Chem. Biol. 2, 228–229.

Ellert, S., Dottir, E., Lenard, A., et al., 2010. Vascular morphogenesis in the zebrafish embryo. Dev. Biol. 341, 56–65.

Eriksson, J., Löfberg, J., 2000. Development of the hypocord and dorsal aorta in the zebrafish embryo (Danio rerio). J. Morphol. 244, 167–176.

Feitsma, H., Cuppen, E., 2008. Zebrafish as a cancer model. Mol. Cancer Res. 6, 685–694.

Haldi, M., Ton, C., Seng, W.L., et al., 2006. Human melanoma cells transplanted into zebrafish proliferate, migrate, produce melanin, form masses and stimulate angiogenesis in zebrafish. Angiogenesis 9, 139–151.

Harfouche, R., Basu, S., Soni, S., et al., 2009. Nanoparticle-mediated targeting of phosphatidylinositol-3-kinase signaling inhibits angiogenesis. Angiogenesis 12, 325–333.

Hasan, J., Shnyder, S.D., Bibby, M., et al., 2004. Quantitative angiogenesis assays in vivo, a review. Angiogenesis 7, 1–16.

Isogai, S., Horiguchi, M., Weisnstein, B.M., 2001. The vascular anatomy of the developing zebrafish: an atlas of embryonic and early larval development. Dev. Biol. 230, 278–301.

Lawson, N.D., Weinstein, B.M., 2002a. Arteries and veins: making a difference with zebrafish. Nat. Rev. Genet. 3, 674–682.

Lawson, N.D., Weinstein, B.M., 2002b. In vivo imaging of embryonic vascular development using transgenic zebrafish. Dev. Biol. 248, 307–318.

Lee, S.L., 2011. The Zebrafish as a Model to Elucidate Human Diseases and Development. Karolinska Institutet, Stockholm.

Lee, S.L., Rouhi, P., Dahl Jensen, L., et al., 2009. Hypoxia-induced pathological angiogenesis mediates tumor cell dissemination, invasion, and metastasis in a zebrafish tumor model. Proc. Natl. Acad. Sci. U.S.A. 106, 19485–19490.

Leong, H.S., Steinmetz, N.F., Ablack, A., et al., 2010. Intravital imaging of embryonic and tumor neovasculature using viral nanoparticles. Nat. Prot. 5, 1406–1417.

Lewis, J.D., Destito, G., Zijlstra, A., et al., 2006. Viral nanoparticles as tools for intravital vascular imaging. Nat. Med. 12, 354–360.

Marques, I.J., Weiss, F.U., Vlecken, D.H., et al., 2009. Metastatic behaviour of primary human tumours in a zebrafish xenotransplantation model. BMC Cancer 9, 128.

Mizgirev, I., Reskoy, S., 2006. Transplantable tumor lines generated in clonal zebrafish. Cancer Res. 66, 3120–3125.

Nasevicius, A., Ekker, S.C., 2000. Effective targeted gene 'knockdown' in zebrafish. Nat. Genet. 26, 216–220.

Nicoli, S., Presta, M., 2007. The zebrafish/tumor xenograft angiogenesis assay. Nat. Protoc. 2, 2918–2923.

Nicoli, S., Ribatti, D., Cotelli, F., et al., 2007. Mammalian tumor xenografts induce neovascularization in zebrafish embryos. Cancer Res. 67, 2927–2931.

Nicoli, S., De Sena, G., Presta, M., 2008a. Fibroblast growth factor 2-induced angiogenesis in zebrafish: the zebrafish yolk membrane (ZFYM) angiogenesis assay. J. Cell. Mol. Med. 13, 2061–2068.

Nicoli, S., Tobia, C., Gualandi, L., et al., 2008b. Calcitonin receptor-like receptor guides arterial differentiation in zebrafish. Blood 111, 4965–4972.

Pichler, F.B., Lauresnson, S., Williams, L.C., et al., 2003. Chemical discovery and global gene expression analysis in zebrafish. Nat. Biotechnol. 21, 879–883.

Rieger, S., Kulkarni, R.P., Darcy, D., et al., 2005. Quantum dots are powerful multipurpose vital labeling agents in zebrafish embryos. Dev. Dyn. 234, 670–681.

Rubinstein, A.L., 2003. Zebrafish: from disease modeling to drug discovery. Curr. Opin. Drug. Discov. Devl. 6, 218–223.

Serbedzija, G.N., Flynn, E., Willett, C.E., 1999. Zebrafish angiogenesis: a new model for drug screening. Angiogenesis 3, 353–359.

Serbedzija, G.N., Flynn, E., Willett, C.E., 2000. Zebrafish angiogenesis: a new model for drug screening. Angiogenesis 3, 353–359.

Snaar-Jagalska, B.E., 2009. ZF-CANCER: developing high-throughput bioassays for human cancers in zebrafish. Zebrafish 6, 441–443.

Spitsbergen, J., 2007. Imaging neoplasia in zebrafish. Nat. Methods 4, 548–549.

Stoletov, K., Klemke, R., 2008. Catch of the day: zebrafish as a human cancer model. Oncogene 27, 4509–4520.

Stoletov, K., Montel, V., Lester, R.D., et al., 2007. High-resolution imaging of the dynamic tumor cell vascular interface in transparent zebrafish. Proc. Natl. Acad. Sci. U.S.A. 104, 17406–17411.

Taylor, A.M., Zon, L.I., 2009. Zebrafish tumor assays: the state of transplantation. Zebrafish 6, 339–346.

Thisse, C., Zon, L.I., 2002. Organogenesis-heart and blood formation from the zebrafish point of view. Science 295, 457–462.

Topczewska, J.M., Postovit, L.M., Margaryan, N.V., et al., 2006. Embryonic and tumorigenic pathways converge via nodal signaling: role in melanoma aggressiveness. Nat. Med. 12, 925–932.

Van Rooijen, E., Voest, E.E., Logister, I., 2010. Von Hippel–Lindau tumor suppressor mutants faithfully model pathological hypoxia-driven angiogenesis and vascular retinopathies in zebrafish. Dis. Model. Mech. 3, 343–353.

Vlecken, D.H., Bagowski, C.P., 2009. LIMK1 and LIMK2 are important for metastatic behavior and tumor cell-induced angiogenesis of pancreatic cancer cells. Zebrafish 6, 433–439.

Weinstein, B., 2002. Vascular cell biology in vivo: a new piscine paradigm? Trend. Cell. Biol. 12, 439–445.

Weinstein, B.M., Stemple, D.L., Driever, W., et al., 1995. A localized heritable vascular patterning defect in the zebrafish. Nat. Med. 1, 1143–1147.

White, J.A., Boffa, M.B., Jones, B., et al., 1994. A zebrafish retinoic acid receptor expressed in the regenerating caudal fin. Development 120, 1861–1872.

White, R.M., Sessa, A., Burke, C., et al., 2008. Transparent adult zebrafish as a tool for in vivo transplantation analysis. Cell Stem Cell 2, 183–189.

Zhang, T.P., Childs, P., Leu, J.P., et al., 2001. Gridlock signaling pathway fashions the first embryonic artery. Nature 414, 216–220.

FURTHER READING

St Croix, B., Rago, C., Velculescu, V., et al., 2000. Genes expressed in human tumor endothelium. Science 289, 1197–1202.

CHAPTER 9

Animal Tumor Models

ANIMAL TUMOR MODELS

In the development of new and more effective antivascular agents, it is of notable importance to have model systems available that can mimic the tumor biology observed in patients and give accurate information about signaling networks involved in vascular maintenance. The study of these clinically relevant parameters required the tumor populations to be placed in the appropriate organ environment, either as orthotopic primary tumor cell inoculum (with subsequent spontaneous metastasis) or as experimental metastasis into relevant organs after intravenous tumor cell injection. Studies have established that the expression of cellular properties required for acquiring both the primary and the metastatic phenotypes is dependent on both intrinsic properties of the tumor cells and host factors, which can differ between tissues and organs (Talmadge et al., 2007).

Numerous animal tumor models have been developed to test the antiangiogenic and anticancer effects of drugs. Subcutaneous (syngeneic) implantation of tumor cells is one of the easiest approaches for investigating tumor growth and angiogenesis in vivo. The classical model of the Lewis lung carcinoma has been extensively used to study antiangiogenic molecules (Fig. 9.1) (O'Reilly et al., 1994). In contrast to subcutaneous models, in orthotopic tumor models (in the organ of origin), the injection of the tumor cells is more technically challenging and tumor take is often low. A third possibility is the xenograft in immunodeficient (nude or severe combined) rodents.

The use of tumor cells transfected to express green fluorescent protein (GFP) enables monitoring of tumor growth (Cross et al., 2008). In this model, the primary tumor and any metastases are detected by an intense green fluorescence, which can be imaged by fluorescence stereomicroscopy and captured on camera (Cross et al., 2008). From high-resolution images of vascular networks, it is possible to generate quantitative measurements including microvascular density and total

In Vivo Models to Study Angiogenesis. DOI: http://dx.doi.org/10.1016/B978-0-12-814020-8.00010-X

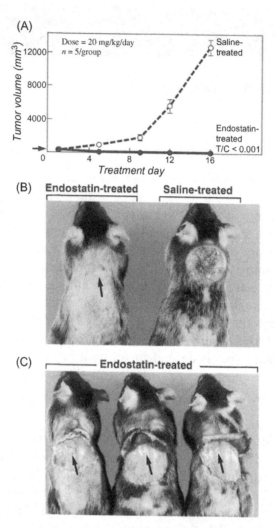

Figure 9.1 Systemic therapy with recombinant endostatin regresses Lewis lung carcinomas primary tumors. The subcutaneous dorsa of mice were implanted with Lewis lung carcinomas. (A) Results of systemic therapy with recombinant mouse endostatin (20 mg/kg/day) initiated when tumors were ≈200 mm³ (1% of body 0.001). Tumors in mice treated with endostatin rapidly regressed and were inhibited by >99% relative to saline-treated controls. (B) Representative treated and control mice after 11 days of systemic therapy with endostatin. Saline-treated mice (right) had rapidly growing red (gray in print versions) tumors with ulcerated surfaces. Endostatin-treated mice (left) had small pale residual tumors (arrow). (C) Residual disease in endostatin-treated mice. Autopsy revealed small white residual tumors at the site of the original primary implantation. Reproduced from O'Reilly, M.S., Boehm, T., Shing, Y., et al., 1997. Endostatin: an endogenous inhibitor of angiogenesis and tumor growth. Cell 88, 277–285.

vessel length, enabling the quantification of angiogenesis, tumor growth, and metastasis (Yang et al., 2001). Okabe et al. (1997) produced transgenic mice expressing GFP in all tissues and these mice have been crossed with nude mice to produce immunodeficient animals, which also have fluorescence green throughout their bodies (Yang et al., 2004). By implanting red fluorescent protein (RFP)-expressing tumors into these mice, it is possible to use whole body imaging to generate dual color images of the early phases of tumor angiogenesis (Yang et al., 2003).

Magnetic resonance imaging (MRI) and computer tomography following the administration of contrast media, or positron emission tomography (PET) visualizing the distribution of a radiolabeled tracer, have been used to visualize mice vasculature and to monitor angiogenesis in drug-treated mice (Rudin and Weissleder, 2003; Neeman and Dafni, 2003).

Fidler (1990) demonstrated that the microenvironment of subcutaneously injected human tumors is radically different from the original milieu and therefore, despite the species differences, the corresponding mouse organ of tumor origin should more closely resemble the original microenvironment than the subcutaneous milieu. Although subcutaneous sites are easily accessible and allow rapid screening of new antitumor compounds, increasing evidence demonstrates that faithful reproduction of the tumor microenvironment and more predictive treatment responses occur with orthotopic tumor implantation, in which either tumor cells or tumor fragments (surgical orthotopic implantation) are implanted within the organ corresponding to the original human tumor site (Fidler, 1991; Hoffman, 1999; Killion et al., 1998).

Orthotopic implantation allows rapid growth of local tumor and spreading of distant metastases and seems to recapitulate the morphology and the growth characteristics of clinical disease (Fidler, 1986, 1991; Hoffman, 1999). Indeed, clinical observation has suggested that the organ environment can influence the response of tumor to chemotherapies (Fidler, 2002; Killion et al., 1998) and orthotopic models allow to obtain a vasculature network similar to that of patient being useful for antivascular drug discovery.

The vasculature of the tumor acquires characteristics similar to those of the host—environment (Langenkamp and Molema, 2009), and the neovasculature of a specific tumor type can be indeed different in terms of vascular architecture, microvascular density, permeability, vessel distribution, and gene expression, when the same tumor is grown in different organs (Langenkamp and Molema, 2009). The microvasculature of human glioblastoma implanted subcutaneously in nude mice became extensively fenestrated, with a large population of caveolae and a relatively high permeability, similar to the host endothelium of the subcutaneous space (Roberts et al., 1998). In contrast, the same tumor implanted in the brain acquired a microvasculature that is considerably less fenestrated, resembling more closely the brain microvascular phenotype. Similarly, human renal carcinoma cells implanted orthotopically into the kidney of nude mice became highly vascularized, whereas the same tumor growing subcutaneously did not (Singh et al., 1994), and the vasculature of melanoma growing intracranially had an higher density, but a smaller diameter, than the vasculature found in subcutaneously growing tumors (Kashiwagi et al., 2005). Differential patterns of expression of angiogenic genes are probably responsible for these host–environment-induced differences in vascularization. Renal cell carcinoma growth in the kidney resulted in a higher expression of fibroblast growth factor.2 (FGF-2), compared with the same tumor growing subcutaneously (Singh et al., 1994). Likewise, the decrease in fenestration pattern and permeability in glioblastoma tumors growing in the brain, compared with those growing subcutaneously, was accompanied by an elevated expression of the receptors for vascular endothelial growth factor (VEGF), whereas expression of VEGF itself did not differ per tumor location (Roberts et al., 1998). Human ductal pancreatic adenocarcinoma grown in the pancreas of nude mice exhibited enhanced expression of VEGF with concomitant higher growth rate compared with subcutaneous tumors implanted in the abdominal wall (Tsuzuki et al., 2001). Colon cancer xenografts in their orthotopic location produced higher levels of interleukin-8 (IL-8) , carcinoembryonic antigen, as well as multidrug resistance proteins, than their subcutaneously growing counterparts (Kitadai et al., 1995). In addition to influence angiogenic genes expression, the host microenvironment can also determine the functionality of genes. Number, nature, and level of active angiogenic genes in a tumor could determine the capacity to overrule the host–environment-driven control of tumor vascular behavior that is intricately controlled by the microenvironment, as demonstrated by Kwei et al. (2004).

These data emphasize the importance of the tumor microenvironment and the active reciprocal communication between tumor cells and the microenvironment, and explain the need for appropriate preclinical animal models representing the clinical setting in which to

Table 9.1 Different Tumors Tested in Orthotopic Model
Colon carcinoma (Thalheimer et al., 2006)
Renal cell carcinoma (Naito et al., 1987)
Breast carcinoma (Price et al., 1990)
Bladder carcinoma (Dinney et al., 1995)
Prostate (Stephenson et al., 1992)
Pancreatic adenocarcinoma (Bruns et al., 1999)
Lung carcinoma (Boehle et al., 2000)
Melanoma (Pastorino et al., 2003a)
Neuroblastoma (Pastorino et al., 2007)

evaluate novel antivascular therapies for cancer in the context of the specific tumor environment. Data from orthotopic tumor systems may give a better indication of the potential clinical activity of antivascular drugs.

The orthotopic implantation of tumor cells into mice is mandatory. The importance of the orthotopic environment for reproducing human tumor conditions has been indeed shown in different tumors (Table 9.1).

TARGETING THE TUMOR VASCULATURE IN VIVO

Small-molecule vascular disrupting agents (VDAs) act on pathophysiological differences between tumor and normal tissue endothelium to achieve selective occlusion of tumor vessels. Among these agents, the combretastatin A-4 (CA-4) is one of the most promising (Lin et al., 1998). The antitumor effects of combretastatin A-4 and its prodrug, the more soluble sodium phosphate salt (CA-4P), has been evaluated in orthotopic models of human colon adenocarcinoma (Grosios et al., 1999) in which tumor growth in the colon is often accompanied by the formation of metastasis in other parts of the body (Cowen et al., 1995). The primary tumor as well as some of the metastatic deposits (including body wall, lymph nodes, and kidney deposits) developed extensive vasculature, underlining the importance of using orthotopic model in antivascular therapeutic studies. Extensive hemorrhagic necrosis after treatments was observed in vascularized but not in avascular tumors in mice at both the primary and metastatic sites. This observation is the clearest evidence that a direct effect of these compounds on tumor vasculature is essential for antitumor activity in vivo.

The effect of CA-4 was also evaluated in combination with Oxi4503, the diphosphate prodrug of CA1P (Pettit et al., 1989) in an orthotopically transplanted human renal cell carcinoma xenograft model, by the vascular casting and the chord-length distribution techniques (Salmon et al., 2006a,b). Results demonstrated the loss of tumor vasculature and induction of wide-scale necrosis in the central regions of the tumor. Compared to the surrounding normal tissue a lower vascular density at the tumor periphery after treatment with either agent was seen, suggesting that invading tumor cells increase the intervascular spacing between normal blood vessels located within this region.

Ligand-targeted VDAs bind selectively to components of tumor blood vessels, carrying the coupled agents to occlude the vessels. Therefore, ligand-directed VDAs are composed of targeting and effector moieties linked together, usually via chemical cross-linkers or peptide bonds (Thorpe, 2004). The targeting moiety is usually an antibody or a peptide directed against markers selectively upregulated on tumor endothelial cells (Thorpe, 2004).

A successful strategy was based on the use of monoclonal antibodies to phosphatidylserine (PS) (Ran et al., 2002). It has been found that PS exposure on vascular endothelial cells is induced by hypoxia/reoxygenation, acidity, as well as by the presence of thrombin, inflammatory cytokines, and reactive oxygen species. This suggests that stress conditions in the tumor microenvironment may be responsible for inducing PS exposure on viable endothelium (Ran and Thorpe, 2002). Ran et al. (2005) tested the use of PS as a potential drug target for breast cancer. By the identification of a monoclonal antibody, 3G4, that binds to PS, they evaluated its effect within orthotopic breast tumor models in severe combined immunodeficiency (SCID) mice. A specific homing to tumor blood vessels and a selective vascular damage with reduction of tumor vascularity and plasma volume, as well as a delay in tumor growth and the development of necrosis were obtained.

Several techniques have been used to find novel markers that could aid the development of targeting moieties with increased tumor specificity over those currently available. In in vivo phage display library screening, a vast numbers of phages, each expressing a different peptide, are injected into animals (Trepel et al., 2002) or terminally ill patients (Arap et al., 2002). A short time later, samples of tumor and normal tissues are removed and phages that have localized into both tumor cells and tumor vasculature are recovered. Phage-contained peptides that confer specific binding are then identified (Trepel et al., 2002).

In vivo phage display-based studies allowed to identify several peptides such as the peptide containing the sequence Arg-Gly-Asp (RGD), an avβ3/avβ5 binding motif, that can bind to the tumor (neo) vasculature (Arap et al., 1998; Pasqualini et al., 1997). Moreover, systemic administration of phages into nude mice led to the selection of a novel tumor vasculature-homing phage carrying the sequence NGR (Arap et al., 1998), and were able to recognize the tumor isoform of aminopeptidase N (APN, CD13). Constitutively, APN is a membrane-bound metallo-peptidase that plays multiple functions as a regulator of various hormones and cytokines, protein degradation, antigen presentation, cell proliferation, cell migration, and angiogenesis (Luan and Xu, 2007; Razak and Newland, 1992; Riemann et al., 1999). This technique allows the identification of the sequence CPRECES, specific for another aminopeptidase, the aminopeptidase A (APA), expressed on endothelial and perivascular tumor cells (Marchiò et al., 2004).

Various compounds, including cytotoxic drugs, cytokines, antiangiogenic agents, viral particles, fluorescent and contrast tracers, DNA complexes, and other biological response modifiers have been coupled or synthetically added to NGR peptides in order to increase their vasculature-homing effects (Arap et al., 1998; Buehler et al., 2006; Corti and Ponzoni, 2004; Curnis et al., 2000, 2005; Grifman et al., 2001; Pastorino et al., 2003a, 2006; Zarovni et al., 2004).

Liposomes are attractive for targeting drugs and other effectors to tumor vasculature because they can carry large payloads. Using nanoparticles of doxorubicin (DOX) targeted to tumor endothelial cells by coupling the liposomes with peptides containing the NGR sequence, and their efficacy in orthotopic model of human neuroblastoma has been evaluated (Pastorino et al., 2003a). Specifically, mice were orthotopically injected in the capsule of the left adrenal gland and tumor growth was evaluated by histological analysis (Fig. 9.2). NGR-targeted, DOX-encapsulated, liposomes caused endothelial and tumor cells apoptosis, and showed antitumor and antimetastatic activity in neuroblastoma (Pastorino et al., 2003a). In a subsequent study, the efficacy of

(A)
Orthotopic neuroblastoma model: injection in the adrenal gland

(1)

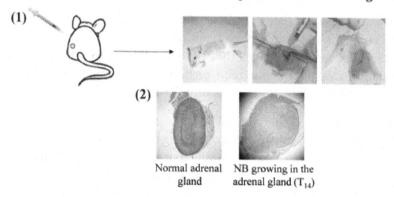

(2)

Normal adrenal gland   NB growing in the adrenal gland (T_{14})

+ metastastatic spreading (liver, spleen, ovary, lung, kidney), T_{21-35}

(B) Tumor growth inhibition by TVT-DOX

(1)

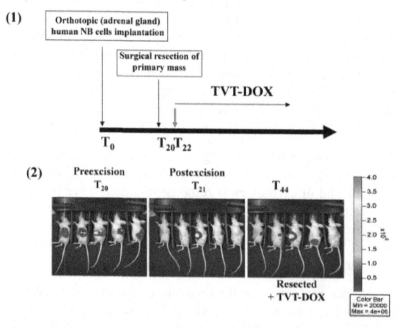

Orthotopic (adrenal gland) human NB cells implantation

Surgical resection of primary mass

TVT-DOX

T_0 $T_{20}T_{22}$

(2) Preexcision T_{20} Postexcision T_{21} T_{44}

Resected + TVT-DOX

Figure 9.2 (A) Neuroblastoma orthotopic implantation procedure. (1) After anesthesia, mice are injected with neuroblastoma cell lines, after laparatomy, in the left adrenal gland. (2) Comparison, by histological analysis, between normal and neuroblastoma-injected adrenal glands 14 days after tumor challenge. The establishment of metastasis is observed in several organs from 21 to 35 days after cell injection. (B) Evaluation of neuroblastoma-tumor growth inhibition by BLI (1) Luciferase-transfected neuroblastoma tumor cells are orthotopically implanted in mice on day 0 and had their tumors surgically resected on day 20. TVT-DOX treatment started at day 22. (2) Orthotopic expansion, achieved tumor resection, and response to TVT-DOX are visualized via a highly sensitive, cooled CCD camera mounted in a lighttight specimen box (IVIS; Xenogen). Animals are injected intraperitoneally with the substrate D-luciferin and are anesthetized with 1%–3% isoflurane and placed onto a warmed stage inside a lighttight camera box, with continuous exposure to 1%–2% isoflurane. Imaging times ranged from 1 s to 3 min depending on the tumor model and the desired time point. The level of bioluminescence is recorded as photons per second. Reproduced from Loi, M., Marchiò, S., Becherini, P., et al., 2011. Combined targeting of perivascular and endothelial tumor cells enhances anti-tumor efficacy of liposomal chemotherapy in neuroblastoma. J. Cont. Release 145, 66–73.

clinical-grade formulation of NGR-peptide-targeted liposomal DOX, TVT-DOX, in several murine orthotopic xenografts of DOX resistant human cancer, including lung, ovarian, and neuroblastoma, has been evaluated (Pastorino et al., 2008). In addition, in order to assess the effect of NGR-targeted liposomal DOX therapy in controlling minimal residual disease (MRD) and in helping to prevent tumor relapse, a new neuroblastoma model was set up to take advantage of new imaging techniques. In vivo bioluminescence imaging (BLI) is indeed a noninvasive assay for the detection of small numbers of cells, and it enables the quantification of tumor growth within internal organs (Edinger et al., 2003). Using a highly sensitive, cooled CCD camera mounted in a lighttight specimen box, the initial trafficking of the malignant cells through the body, and organ-specific homing and orthotopic expansion over time were readily visualized and quantified. A reporter gene that codes for a bioluminescent marker to permanently label neuroblastoma cell lines was used, and primary tumors and metastases growth following orthotopic implantation into mice were analyzed. Treatment with TVT-DOX induced a partial arrest in primary tumor growth and an inhibition of MRD in treated mice. Moreover, TVT-DOX resulted in effectively killing angiogenic tumor blood vessels and, indirectly, the tumor cells that these vessels supported (Pastorino et al., 2008).

On the basis of the "marker heterogeneity" concept, pointed out by Kerbel et al. (2001), tumors can modulate the markers they induce on the adjacent endothelia, giving rise to heterogeneity in the expression of tumor vessel markers. The use of combinations of VDAs that recognize two, or more, differently regulated, tumor vessel markers could enhance the therapeutic effects. A dual vascular targeting efficacy was evaluated in orthotopically implanted tumor animal models by targeting both the endothelial cells and pericytes (Lu et al., 2007). The combination therapy showed a significantly decrease in pericyte coverage with reduction of microvascular density and an enhanced endothelial cell apoptosis.

With the similar purpose, neuroblastoma orthotopic tumors have been studied for the therapeutic efficacy of a novel, APA-targeted, liposomal formulation of DOX, alone or in combination with APN-targeted DOX-loaded liposomes. Results showed that the use of APA- and APN-targeted, DOX-entrapped, liposomes administered in combination led to a significant increase in life span compared to each treatment

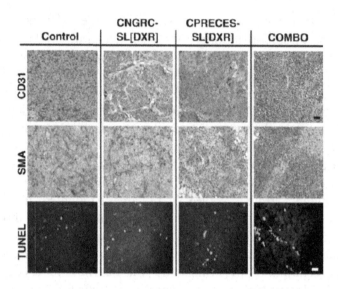

Figure 9.3 Effects of APN- and APA-targeted SL[DXR] combination on endothelial, perivascular, and tumor cells in vivo. Immunohistochemistry was performed on established neuroblastoma tumors removed from untreated mice (control) or from mice treated with DXR-loaded, APN-targeted, or APA-targeted liposomes, or with a combination of the two liposomal formulations (COMBO). Tumors were removed on day 36 and tissue sections were immunostained for CD31 and SMA to detect tumor vasculature. TUNEL was performed to detect tumor apoptosis. Cell nuclei were stained with DAPI. Reproduced from Loi, M., Marchiò, S., Becherini, P., et al. 2011. Combined targeting of perivascular and endothelial tumor cells enhances anti-tumor efficacy of liposomal chemotherapy in neuroblastoma. J. Cont. Release 145, 66–73.

administered separately, obtaining a strong enhanced antitumor effect and induction of endothelial cells apoptosis and decrease in pericyte coverage (Fig. 9.3) (Loi et al., 2010).

COMBINATION THERAPY INVOLVING VDAS

VDAs selectively destroy the central region of tumors leaving a peripheral rim of survival tumor cells (Thorpe et al., 2003, Thorpe, 2004). VDAs are most effective against vessels in the interior of the tumor due to the high interstitial pressure in these regions and the vascular architecture. Interstitial fluid pressure rises from the tumor periphery to the tumor center (Boucher et al., 1990), so that an increase in vascular permeability is tolerated only at the periphery (Tozer et al., 2005). Small-caliber vessels are also more sensitive to shutdown than larger ones and the proportion of these is often far higher at the center than at the periphery (Tozer et al., 1999). There is often a complex vascular plexus at the tumor periphery, compared with a much lower vascular density at the center, so that in the event of extensive vascular damage, residual flow is likely to persist at the periphery rather than at the center. In this context, cells in this rim of viable tumor tissue are likely to be highly proliferative and well nourished. These characteristics, coupled with the accessibility of the surviving tumor cells to systemically administered agents, make these cells susceptible to killing by conventional methods. Conversely, VDAs induce the destruction of large areas of the tumor center and exhibit excellent activity against bulky disease (Landuyt et al., 2001; Siemann and Shi, 2003), which is typically resistant to conventional anticancer treatment. Thus, combining VDAs with cytotoxic chemotherapy, radiation and antiangiogenic treatments would be expected to lead to additive or even synergistic antitumor activity (Siemann et al., 2002; Thorpe, 2004).

The combination of two different targeting strategies was investigated. Effect of the combination therapy was investigated using the monoclonal antibody 3G4 and chemotherapeutic agents, such as gemcitabine and docetaxel, in pancreatic and breast orthotopic models, respectively (Beck et al., 2006; Huang et al., 2005). The effect of the combination was significantly better than either therapy alone, suggesting that gemcitabine enhances 3G4 function, or vice versa (Beck et al., 2006). Similarly, 3G4 enhanced the inhibitory effect of docetaxel on the growth of orthotopically implanted breast tumor model, pointing out that tumor microenvironment plays an important role in docetaxel-induced exposure of anionic phospholipids on tumor vascular endothelium (Huang et al., 2005).

The murine anti-PS antibody, 2aG4, in combination with local irradiation, was used to treat established gliomas growing in the brain of syngeneic rats (He et al., 2009). 2aG4 treatment not only has an antivascular action when combined with irradiation in this model, but also enhances the immunogenicity of the tumor leading to immunological control of residual tumor cells. Moreover, antivascular effects and antitumor effects were accompanied by the recruitment of host immune cells in tumor vasculature and the subsequent infiltration into the tumor interstitium (Beck et al., 2006; Tozer et al., 2008).

A successful combination strategy in neuroblastoma model using liposomal formulations of DXR targeted against both tumor cells, via anti-GD2 monoclonal antibodies, and against the tumor vasculature, via the NGR peptide (Pastorino et al., 2006). The anti-GD2-targeted

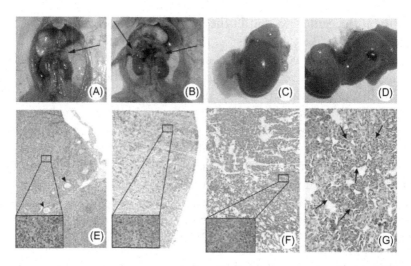

Figure 9.4 Orthotopic neuroblastoma xenograft model in SCID mice. (A and B) Adrenal gland tumors (arrows) in mice that were injected orthotopically with SH-SY5Y cells at 14 (A) and 21 (B) days before sacrifice. (C and D) Representative right adrenal gland (C) and liver (D) samples at 3 and 4 weeks after injection of neuroblastoma cells, respectively. (E–H) Histological analysis of representative ovary (E), kidney (F), liver (G), and lung (H) samples. Forty days after cell injection, animals were sacrificed, the organs were removed, fixed, paraffin embedded, sectioned at 5 μm, and stained with hematoxylin & eosin. Arrows indicate metastatic tumor invasion in the lung. Arrowheads show the normal ovaric follicular structure surrounded by tumor neuroblastoma cells. Reproduced from Pastorino, F., Brignole, C., Marimpietri, D., et al., 2003a. Vascular damage and anti-angiogenic effects of tumor vessel-targeted liposomal chemotherapy. *Cancer Res.* 63, 7400–7409.

liposomes resulted in direct cell kill, including cytotoxicity against cells that were at the tumor periphery and were independent of the tumor vasculature, whereas NGR-peptide-targeted liposomal DOX bind to and killed angiogenic blood vessels and, indirectly, the tumor cells that these vessels supported, mainly in the tumor core. The results clearly showed that liposomes administered in a sequential manner were statistically more effective in inhibiting neuroblastoma-tumor proliferation in mice compared with formulations given alone (Fig. 9.4) (Pastorino et al., 2003a,b).

TRANSGENIC MOUSE MODELS OF TUMOR ANGIOGENESIS

In these models, animals overexpress targeted oncogenes leading to spontaneous tumor formation over a longer time period. These include the RIP1-TAG2 transgenic mice, the transgenic mice carrying the BPV-1 oncogene, the keratin-14 (K14)-human papilloma virus-16 (HPV16), and the papilloma virus type 1 transgenic tumor models (Fig. 9.5).

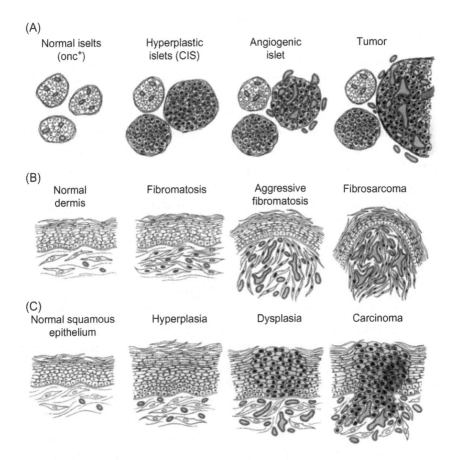

Figure 9.5 The angiogenic switch occurs prior to tumor formation in three transgenic mouse models of tumorigenesis. (A) Expression of the Tag oncogene in the pancreatic islets elicits four sequential stages in tumor development: normal, oncogene-expressing islets; hyperplastic islets, populated by proliferating; angiogenic islets; and solid tumors. (B) In transgenic mice carrying the BPV-1 oncogenes, the normal dermis is initially converted into a state of mild fibromatosis, revealed as focal accumulation of dermal fibroblasts. Angiogenesis is first evident in the next stage, aggressive fibromatosis, which is also marked by dense arrays of proliferating fibroblastic cells; both hyperproliferation and angiogenesis persist in the subsequent stage, protuberant fibrosarcoma. (C) Targeted expression of the HPV-16 oncogenes to basal cells of the epidermis induces multistage development of squamous cell carcinoma, beginning as hyperplasia of keratinocytes, which progresses to dysplasia, marked by morphologically aberrant keratinocytes with a high proliferation index and by abundant neovascularization; finally, two classes of squamous carcinoma, with extensive angiogenesis. Reproduced from Hanahan, D., Folkman, J., 1986. Patterns and emerging mechanisms of the angiogenic switch during tumorigenesis. Cell 86, 353−364.

THE RIP1-TAG2 MODEL AND THE TRANSGENIC MICE CARRYING THE BPV-1 ONCOGENES

Hanahan (1985) developed transgenic mice in which the large T oncogene is hybridized to the insulin promoter. In this islet cell tumorigenesis, these mice express the large T antigen in all their islet cells at birth, and express the SV40 T antigen (TAG) under the control of the insulin gene promoter, which elicits the sequential development of tumors in the islets over a period of 12–14 weeks. Tumor development proceeds by stages during which about half the 400 islets become hyperproliferate, while a subset (about 25%) subsequently acquires the ability to switch to angiogenesis. Some 15%–20% of these angiogenic islets develop into benign tumors, encapsulated lesions, and invasive carcinomas. This multistage pathway suggests the sequential involvement of multiple rate-limiting genetic and epigenetic events in the progression from normal cells to tumors. The ß cells become hyperplastic and progress to tumors via a reproducible and predictable multistep process (Hanahan, 1985). One step occurs at 6–7 weeks, when angiogenesis is switched on in approximately 10% of preneoplastic islets. Solid vascularized tumors first appear at 9–10 weeks, initially as small nodules that grow and progress to large islet tumors, with well-defined margins, as well as two classes of invasive carcinoma (Lopez, Hanahan, 2002). Lopez and Hanahan identified stage-specific molecular markers accessible via the circulation, either on the surface of endothelial cells, their peri-endothelial support cells (pericytes and smooth muscle cells), or even tumor cells themselves (as a result of the hemorrhagic leaky angiogenic vasculature). They selected phage pools that homed preferentially to different stages during RIP1-TAG2 tumorigenesis. In addition to angiogenic markers shared by many types of tumors, they identified vascular target molecules characteristic of this tumor's tissue of origin and not expressed in the vessels of several tumor types growing in or under the skin. Two concepts emerged from this early characterization of tumorigenesis in RIP-TAG transgenic mice: (1) the existence of distinct stages of premalignant progression, namely a hyperplastic stage followed by a stochastic angiogenic stage and (2) the development of angiogenesis well before the emergence of an invasive malignancy. The temporal and histological changes that occur in the RIP-TAG model are consistent with the multistep paradigm for tumorigenesis of human cancers (Vogelstein, 1993). The high incidence of occult human cancers suggests that this angiogenic switch may, as in the RIP-TAG model, be a relatively late event that plays a significant role in the transition from microscopic foci to macroscopic tumor (Udagawa et al., 2002). Bergers et al. (2000) demonstrated that matrix metallopeptidase-9 (MMP-9) plays a crucial role in the initial angiogenic switch during islet carcinogenesis and proposal mobilization of VEGF from an extracellular reservoir as its mode of action.

Joyce et al. (2004) have shown that a subset of papain family Clan CA proteases known as cathepsins make an important contribution to the development of islet tumors and are upregulated during their progression. Increased activity was associated with the angiogenic vasculature and invasive fronts of carcinomas. A broad-spectrum cysteine inhibitor that knocked out cathepsin function at different stages of tumorigenesis impaired angiogenic switching in progenitor lesions, as well as tumor growth, vascularity, and invasiveness. Cysteine cathepsins are also upregulated during HPV16-induced cervical carcinomas. Joyce et al. (2003, 2005) have since shown that heparanase expression increases during RIP-TAG tumorigenesis, predominantly supplied by innate immune cell infiltrating neoplastic tissues, and analyzed the vasculature in the angiogenic stages of RIP-TAG model islet tumorigenesis with phage libraries that display short peptides, and identified peptides that discriminate between the vasculature of the premalignant angiogenic islets and the fully developed vasculature. One peptide is homologous with PDGF-B, which is expressed in endothelial cells, while its receptor is expressed in pericytes. Three PDGF ligand genes are expressed in the tumor endothelial cells, while PDGFB receptor (PDGFB-R) is expressed in tumor pericytes (Bergers et al., 2003). RIP-TAG transgenic mice provides a useful model for examining the effects of various antiangiogenic and antitumorigenic agents, as it allows these compounds to be tested in a stage-specific manner (Bergers et al., 1999).

THE PAPILLOMA VIRUS TYPE 1 TRANSGENIC TUMOR MODEL

In this model, formation of dermal fibrosarcomas in BPV1.69 transgenic mice occurred in three histologically distinct stages (normal, mild, and aggressive fibromatosis) (Sippola-Thiele et al., 1988) characterized by differential expression of c-jun and jun-b proto-oncogenes and their associated AP1 transcription factor activities (Bossy-Wtzel et al., 1992). Evaluation of microvascular density revealed a dramatic increase in capillaries in the aggressive fibromatosis stage, and conditioned medium from cells derived from normal dermis and mild fibromatosis did not display endothelial cell mitogenic activity, whereas those from the aggressive fibromatosis and fibrosarcoma did (Kandell et al., 1991).

THE K14-HPV16 TRANSGENIC TUMOR MODEL

The first pattern of upregulation of angiogenesis-inducer genes is evident during epidermal squamous carcinogenesis in K14-HPV16 transgenic mice, expressing HPV type 16 oncogene under control of the K14 promoter. These mice express the HPV16 E6 and E7 oncogenes in the basal cells of their squamous epithelia (Arbei et al., 1994) in the FVB/n strain background.

They spontaneously develop epidermal squamous cell cancers (SCC) in a multistage fashion (Coussens et al., 1996). Their skin appears normal at birth, but becomes hyperplastic within the first month, and focal dysplasias develop between 3 and 6 months of age. These focal dysplasias are angiogenic and by 1 year have developed into invasive SCC in about half of the mice. Both well and moderately differentiated SCC arise from pathways beginning as hyperplasia and progressing through varying degrees of dysplasia. A perceptible increase in dermal capillary density is first apparent in the first-month hyperplastic stage. There is a striking increase in both the number and distribution of dermal capillaries in the early and advanced dysplastic lesions; numerous vessels become closely apposed to the basement membrane separating dysplastic keratinocytes from the underlying stroma. The pattern is indicative of an angiogenic switch from vascular quiescence to the modest angiogenesis seen in the early, low-grade lesions, followed by a second, striking upregulation of angiogenesis in high-grade neoplasias as well as invasive cancer.

Progression is accompanied by the upregulation of proangiogenic factors, such as VEGF and FGF-2, and the model has called attention to the involvement of proteases from inflammatory mast cells, neutrophils, and macrophages in angiogenesis and tumor progression (Coussens et al., 1999, 2000; Arbeit et al., 1996; Smith-Mc Cune et al., 1997).

Arbeit et al. (1994) elaborated a derivative model of estrogen-induced carcinoma in female K14-HPV16 mice. Cervical carcinoma developed in several stages in 80% of these mice after 6 months of estrogen treatment. Neither the oncogene SV40TAG used in the creation of the RIP-TAG mice nor oncogenes E6 and E7 used in the creation of K14-HPV16 mice induced angiogenesis on their own. Additional changes during progression enable angiogenesis.

Giraudo et al. (2004) used the K14-HPV16 transgenic tumor model to demonstrate MMP-9 in the tumor stroma concomitant with the angiogenic switch, expressed by infiltrating macrophages. Moreover, preclinical trials targeting MMP-9 and angiogenesis with an MMP inhibitor and with a bisphosphonate, zoledronic acid, showed that both were antiangiogenic. Other enzymes are involved in islet tumorigenesis.

Advantages

1. The orthotopic model reproducing the appropriate tumor microenvironment allows the emergence of the biological features of cancer progression, angiogenic process, metastatic phenotype, resistance, and therapeutic response to therapies with similar patterns observed in human cancer and, for these reasons, appears appropriate for studying antivascular approaches.
2. It is possible to use such models to determine whether a new drug is antiangiogenic or antivascular in action. If the former, the drug will prevent or reduce the growth of new blood vessels to the tumor; if the latter, it will damage the endothelial lining of the existing tumor blood vessels.
3. MRI and PET tracers may be used to monitor angiogenesis in drug-treated mice.

Disadvantage

1. The tumors are established within a few weeks after cell implantation, whereas human cancer develops over a period of several months or years.

REFERENCES

Arap, W., Pasqualini, R., Ruoslahti, E., 1998. Cancer treatment by targeted drug delivery to tumor vasculature in a mouse model. Science 279, 377–380.

Arap, W., Kolonin, M.G., Trepel, M., et al., 2002. Steps toward mapping the human vasculature by phage display. Nat. Med 8, 121–127.

Arbeit, J., Munger, K., Howley, P., et al., 1994. Progressive squamous epithelial neoplasia in K14-HPV16 transgenic mice. J. Virol. 68, 4358–4368.

Arbeit, J., Olson, D., Hanahan, D., 1996. Upregulation of fibroblast growth factors and their receptors during multistage epidermal carcinogenesis in K14-HPV16 transgenic mice. Oncogene 13, 1847–1857.

Beck, A.W., Luster, T.A., Miller, A.F., et al., 2006. Combination of a monoclonal anti-phosphatidylserine antibody with gemcitabine strongly inhibits the growth and metastasis of orthotopic pancreatic tumors in mice. Int. J. Cancer 118, 2639–2643.

Bergers, G., Javaherian, K., Lo, K., et al., 1999. Effects of angiogenesis inhibitors on multistage carcinogenesis in mice. Science 284, 808–812.

Bergers, G., Brekken, R., Mc Mahon, J., et al., 2000. Matrix metalloproteinase-9 triggers the angiogenic switch during carcinogenesis. Nat. Cell. Biol. 2, 737–744.

Bergers, G., Song, S., Meyer-Morse, N., et al., 2003. Benefits of targeting both pericytes and endothelial cells in the tumor vasculature with kinase inhibitors. J. Clin. Invest. 111, 1287–1295.

Boehle, A.S., Dohrmann, P., Leuschner, I., et al., 2000. An improved orthotopic xenotransplant procedure for human lung cancer in SCID bg mice. Ann. Thorac. Surg. 69, 1010–1015.

Bossy-Wtzel, E., Bravo, R., Hanahan, D., 1992. Transcription factors JunB and cJun are selectively up-regulated and functionally implicated in fibrosarcoma development. Genes Dev. 6, 2340–2351.

Boucher, Y., Baxter, L.T., Jain, R.K., 1990. Interstitial pressure gradients in tissue-isolated and subcutaneous tumors: implications for therapy. Cancer Res. 50, 4478–4484.

Bruns, C.J., Harbison, M.T., Kuniyasu, H., et al., 1999. In vivo selection and characterization of metastatic variants from human pancreatic adenocarcinoma by using orthotopic implantation in nude mice. Neoplasia 1, 50–62.

Buehler, A., Van Zandvort, M.A., Stelt, B.J., et al., 2006. cNGR: a novel homing sequence for CD13/APN targeted molecular imaging of murine cardiac angiogenesis in vivo. Arterioscler. Thromb. Vasc. Biol. 26, 2681–2687.

Corti, A., Ponzoni, M., 2004. Tumor vascular targeting with tumor necrosis factor alpha and chemotherapeutic drugs. Ann. N.Y. Acad. Sci. 1028, 104–112.

Coussens, L., Raymond, W., Bergers, G., et al., 1999. Inflammatory mast cells up-regulate angiogenesis during squamous epithelial carcinogenesis. Genes Dev. 13, 1382–1397.

Coussens, L., Tinkle, C., Hanahan, D., et al., 2000. MMP-9 supplied by bone marrow-derived cells contribute to skin carcinogenesis. Cell 103, 481–490.

Coussens, L.M., Hanahan, D., Arbeit, J., 1996. Genetic predisposition and parameters of malignant progression in K14-HPV16 transgenic mice. Am. J. Pathol. 149, 1899–1917.

Cowen, S.E., Bibb, M.C., Double, J.A., 1995. Characterization of the vasculature within a murine adenocarcinoma growing in different sites to evaluate the potential of vascular therapies. Acta Oncol. 34, 357–360.

Cross, N.A., Fowles, A., Reeves, K., et al., 2008. Imaging the effects of castration on bone turnover and hormone-independent prostate cancer colonization of bone. Prostate. 68, 1707–1714.

Curnis, F., Sacchi, A., Borgna, L., et al., 2000. Enhancement of tumor necrosis factor alpha antitumor immunotherapeutic properties by targeted delivery to aminopeptidase N (CD13). Nat. Biotechnol. 18, 1185–1190.

Curnis, F., Gasparri, A., Sacchi, A., et al., 2005. Targeted delivery of IFNgamma to tumor vessels uncouples antitumor from counterregulatory mechanisms. Cancer Res. 65, 2906–2913.

Dinney, C.P., Fishbeck, R., Singh, R.K., et al., 1995. Isolation and characterization of metastatic variants from human transitional cell carcinoma passaged by orthotopic implantation in athymic nude mice. J. Urol. 154, 1532–1538.

Edinger, M., Cao, Y.A., Verneris, M.R., et al., 2003. Revealing lymphoma growth and the efficacy of immune cell therapies using in vivo bioluminescence imaging. Blood 101, 640–648.

Fidler, I.J., 1986. Rationale and methods for the use of nude mice to study the biology and therapy of human cancer metastasis. Cancer Metastasis Rev 5, 29–49.

Fidler, I.J., 1990. Critical factors in the biology of human cancer metastasis: twenty-eighth G.H. A. Clowes memorial award lecture. Cancer Res. 50, 6130–6138.

Fidler, I.J., 1991. Orthotopic implantation of human colon carcinomas into nude mice provides a valuable model for the biology and therapy of metastasis. Cancer Metastasis Rev 10, 229–243.

Fidler, I.J., 2002. The organ microenvironment and cancer metastasis. Differentiation 70, 498–505.

Giraudo, E., Inoue, M., Hanahan, D., 2004. An amino-bisphosphonate targets MMP-9-expressing macrophages and angiogenesis to impair cervical carcinogenesis. J. Clin. Invest. 114, 623–633.

Grifman, M., Trepel, M., Speece, P., et al., 2001. Incorporation of tumor-targeting peptides into recombinant adeno-associated virus capsids. Mol. Ther. 3, 964–975.

Grosios, K., Holwell, S.E., Mcgown, A.T., et al., 1999. In vivo and in vitro evaluation of combretastatin A-4 and its sodium phosphate prodrug. Br. J. Cancer 81, 1318–1327.

Hanahan, D., 1985. Heritable formation of pancreatic beta-cell tumours in transgenic mice expressing recombinant insulin/simian virus 40 oncogenes. Nature 315, 115–122.

He, J., Yin, Y., Luster, T.A., et al., 2009. Antiphosphatidylserine antibody combined with irradiation damages tumor blood vessels and induces tumor immunity in a rat model of glioblastoma. Clin. Cancer Res. 15, 6871–6880.

Hoffman, R.M., 1999. Orthotopic metastatic mouse models for anticancer drug discovery and evaluation: a bridge to the clinic. Invest. New Drugs 17, 343–359.

Huang, X., Bennett, M., Thorpe, P.E., 2005. A monoclonal antibody that binds anionic phospholipids on tumor blood vessels enhances the antitumor effect of docetaxel on human breast tumors in mice. Cancer Res. 65, 4408–4416.

Joyce, J.A., Laakkonen, P., Bernasconi, M., et al., 2003. Stage-specific markers revealed by phage display in a mouse model of pancreatic islet tumorigenesis. Cancer Cell 4, 393–403.

Joyce, J.A., Baruk, A., Kareem Chehade, K., et al., 2004. Cathepsin cysteine proteases are effectors of invasive growth and angiogenesis during multistage tumorigenesis. Cancer Cell 5, 443–453.

Joyce, J.A., Freeman, C., Meyer-Morse, N., et al., 2005. A functional mimetic implicates heparanase and its target heparin sulfate in tumor angiogenesis and invasion in a mouse model of multistage cancer. Oncogene 24, 4037–4051.

Kandel, J., Bossy-Wetzel, E., Radvanyi, F., et al., 1991. Neovascularization is associated with a switch to the export of bFGF in the multistep development of fibrosarcoma. Cell 66, 1095–1104.

Kerbel, R.S., Yu, J., Tran, J., et al., 2001. Possible mechanisms of acquired resistance to anti-angiogenic drugs: implications for the use of combination therapy approaches. Cancer Metastasis Rev 20, 79–86.

Killion, J.J., Radinsky, R., Fidler, I.J., 1998. Orthotopic models are necessary to predict therapy of transplantable tumors in mice. Cancer Metastasis Rev 17, 279–284.

Kitadai, Y., Bucana, C.D., Ellis, L.M., et al., 1995. In situ mRNA hybridization technique for analysis of metastasis-related genes in human colon carcinoma cells. Am. J. Pathol. 147, 1238–1247.

Kwei, S., Stavrakis, G., Takahas, M., et al., 2004. Early adaptive responses of the vascular wall during venous arterialization in mice. Am. J. Pathol. 164, 81–89.

Landuyt, W., Ahmed, B., Nuyts, S., et al., 2001. In vivo antitumor effect of vascular targeting combined with either ionizing radiation or anti-angiogenesis treatment. Int. J. Radiat. Oncol. Biol. Phys. 49, 443–450.

Langenkamp, E., Molema, G., 2009. Microvascular endothelial cell heterogeneity: general concepts and pharmacological consequences for anti-angiogenic therapy of cancer. Cell Tissue Res 335, 205–222.

Lin, C.M., Singh, S.B., Chu, P.S., et al., 1998. Interactions of tubulin with potent natural and synthetic analogs of the antimitotic agent combretastatin: a structure–activity study. Mol. Pharmacol. 34, 200–208.

Loi, M., Marchiò, S., Becherini, P., et al., 2010. Combined targeting of perivascular and endothelial tumor cells enhances anti-tumor efficacy of liposomal chemotherapy in neuroblastoma. J. Cont. Release 145, 66–73.

Lopez, T., Hanahan, D., 2002. Elevated levels of IGF-1 receptor convey invasive and metastatic capability in a mouse model of pancreatic islet tumorigenesis. Cancer Cell 1, 339–353.

Lu, C., Kamat, A.A., Lin, Y.G., et al., 2007. Dual targeting of endothelial cells and pericytes in antivascular therapy for ovarian carcinoma. Clin. Cancer Res. 13, 4209–4217.

Luan, Y., Xu, W., 2007. The structure and main functions of aminopeptidase N. Curr. Med. Chem. 14, 639–647.

Marchiò, S., Lahdenranta, J., Schlingemann, R.O., et al., 2004. Aminopeptidase A is a functional target in angiogenic blood vessels. Cancer Cell. 5, 151–162.

Naito, S., Von Eschenbach, A.C., Fidler, I.J., 1987. Different growth pattern and biologic behavior of human renal cell carcinoma implanted into different organs of nude mice. J. Natl. Cancer Inst. 78, 377–385.

Neeman, M., Dafni, H., 2003. Structural, functional, and molecular MR imaging of the microvasculature. Preclinical MRI experience in imaging angiogenesis. Annu. Rev. Biomed. Eng. 5, 29–56.

Okabe, M., Ikawa, M., Kominami, K., et al., 1997. Green mice" as a source of ubiquitous green cells. FEBS Lett 407, 313–319.

O'Reilly, M.S., Holmgren, L., Shing, Y., et al., 1994. Angiostatin: a novel angiogenesis inhibitor that mediates the suppression of metastases by a Lewis lung carcinoma. Cell 79, 315–328.

Pastorino, F., Brignole, C., Marimpietri, D., et al., 2003a. Vascular damage and anti-angiogenic effects of tumor vessel-targeted liposomal chemotherapy. Cancer Res. 63, 7400–7409.

Pastorino, F., Brignole, C., Marimpietri, D., et al., 2003b. Doxorubicin-loaded Fab' fragments of anti-disialoganglioside immunoliposomes selectively inhibit the growth and dissemination of human neuroblastoma in nude mice. Cancer Res. 63, 86–92.

Pastorino, F., Brignole, C., Di Paolo, D., et al., 2006. Targeting liposomal chemotherapy via both tumor cell-specific and tumor vasculature-specific ligands potentiates therapeutic efficacy. Cancer Res. 66, 10073–10082.

Pastorino, F., Marimpietri, D., Brignole, C., et al., 2007. Ligand-targeted liposomal therapies of neuroblastoma. Curr. Med. Chem. 14, 3070–3078.

Pastorino, F., Di Paolo, D., Piccardi, F., et al., 2008. Enhanced antitumor efficacy of clinical-grade vasculature-targeted liposomal doxorubicin. Clin. Cancer Res. 14, 7320–7329.

Pettit, G.R., Singh, S.B., Hamel, E., et al., 1989. Isolation and structure of the strong cell growth and tubulin inhibitor combretastatin A-4. Experientia 45, 209–211.

Price, J.E., Polyzos, A., Zhang, R.D., et al., 1990. Tumorigenicity and metastasis of human breast carcinoma cell lines in nude mice. Cancer Res. 50, 717–721.

Ran, S., Thorpe, P.E., 2002. Phosphatidylserine is a marker of tumor vasculature and a potential target for cancer imaging and therapy. Int. J. Radiat. Oncol. Biol. Phys. 54, 1479–1484.

Ran, S., He, J., Huang, X., et al., 2005. Antitumor effects of a monoclonal antibody that binds anionic phospholipids on the surface of tumor blood vessels in mice. Clin. Cancer Res. 11, 1551–1562.

Ran, S., Downes, A., Thorpe, P.E., 2002. Increased exposure of anionic phospholipids on the surface of tumor blood vessels. Cancer Res. 62, 6132–6140.

Razak, K., Newland, A.C., 1992. The significance of aminopeptidases and haematopoietic cell differentiation. Blood Rev. 6, 243–250.

Riemann, D., Kehlen, A., Langner, J., 1999. CD13 not just a marker in leukemia typing. Immunol. Today 20, 83–88.

Roberts, W.G., Delaat, J., Nagane, M., et al., 1998. Host microvasculature influence on tumor vascular morphology and endothelial gene expression. Am. J. Pathol. 153, 1239–1248.

Rudin, M., Weissleder, R., 2003. Molecular imaging in drug discovery and development. Nat. Rev. Drug. Discov. 2, 123–131.

Salmon, H.W., Mladinich, C., Siemann, D.W., 2006a. Evaluations of the renal cell carcinoma model Caki-1 using a silicon based microvascular casting technique. Technol. Cancer Res. Treat. 5, 45–51.

Salmon, H.W., Mladinich, C., Siemann, D.W., 2006b. Evaluations of vascular disrupting agents CA4P and OXi4503 in renal cell carcinoma (Caki-1) using a silicon based microvascular casting technique. Eur. J. Cancer 42, 3073–3078.

Siemann, D.W., Shi, W., 2003. Targeting the tumor blood vessel network to enhance the efficacy of radiation therapy. Semin. Radiat. Oncol. 13, 53–61.

Siemann, D.W., Mercer, E., Lepler, S., et al., 2002. Vascular targeting agents enhance chemotherapeutic agent activities in solid tumor therapy. Int. J. Cancer 99, 1–6.

Singh, R.K., Bucana, C.D., Gutman, M., et al., 1994. Organ site-dependent expression of basic fibroblast growth factor in human renal cell carcinoma cells. Am. J. Pathol. 145, 365–374.

Sippola-Thiele, M., Hanahan, D., Howley, P.M., 1988. Cell heritable stages of tumor progression in transgenic mice harboring the bovine papilloma virus type 1 genome. Mol. Cell Biol. 9, 925–934.

Smith-Mc Cune, K., Zhu, Y.H., Hanahan, D., et al., 1997. Cross-species comparisons of angiogenesis during the premalignant stages of squamous carcinogenesis in the human cervix and K14-HOV16 transgenic mice. Cancer Res. 57, 1294–1300.

Stephenson, R.A., Dinney, C.P., Gohji, K., et al., 1992. Metastatic model for human prostate cancer using orthotopic implantation in nude mice. J. Natl. Cancer Inst. 84, 951–957.

Talmadge, J.E., Singh, R.K., Fidler, I.J., et al., 2007. Murine models to evaluate novel and conventional therapeutic strategies for cancer. Am. J. Pathol. 170, 793–804.

Thalheimer, A., Illert, B., Bueter, M., et al., 2006. Feasibility and limits of an orthotopic human colon cancer model in nude mice. Comp. Med 56, 105–109.

Thorpe, P.E., 2004. Vascular targeting agents as cancer therapeutics. Clin. Cancer Res. 10, 415–427.

Thorpe, P.E., Chaplin, D.J., Blakey, D.C., 2003. The first international conference on vascular targeting: meeting overview. Cancer Res. 63, 1144–1147.

Tozer, G.M., Prise, V.E., Wilson, J., et al., 1999. Combretastatin A-4 phosphate as a tumor vascular-targeting agent: early effects in tumors and normal tissues. Cancer Res. 59, 1626–1634.

Tozer, G.M., Kanthou, C., Baguley, B.C., 2005. Disrupting tumour blood vessels. Nat. Rev. Cancer 5, 423–435.

Trepel, M., Arap, W., Pasqualini, R., 2002. In vivo phage display and vascular heterogeneity: implications for targeted medicine. Curr. Opin. Chem. Biol. 6, 399–404.

Tsuzuki, Y., Mouta Carreira, C., et al., 2001. Pancreas microenvironment promotes VEGF expression and tumor growth: novel window models for pancreatic tumor angiogenesis and microcirculation. Lab. Invest. 81, 1439–1451.

Udagawa, T., Fernandez, A., Achilles, E.G., et al., 2002. Persistence of microscopic human cancers in mice: alterations in the angiogenic balance accompanies loss of tumor dormancy. FASEB J 16, 1361–1370.

Vogelstein, B., 1993. The multistep nature of cancer. Trend. Genet 9, 138–141.

Yang, M., Baranov, E., Li, X.M., et al., 2001. Whole-body and intravital optical imaging of angiogenesis in orthotopically implanted tumors. Proc. Natl. Acad. Sci. U.S.A. 98, 2616–2621.

Yang, M., Li, L., Jiang, P., et al., 2003. Dual-color fluorescence imaging distinguishes tumor cells from induced host angiogenic vessels and stromal cells. Proc. Natl. Acad. Sci. U.S.A. 100, 14259–14262.

Yang, M., Reynoso, J., Jiang, P., et al., 2004. Transgenic nude mouse with ubiquitous green fluorescent protein expression as a host for human tumors. Cancer Res. 64, 8651–8656.

Zarovni, N., Monaco, L., Corti, A., 2004. Inhibition of tumor growth by intramuscular injection of cDNA encoding tumor necrosis factor alpha coupled to NGR and RGD tumor-homing peptides. Hum. Gene Ther. 15, 373–382.

FURTHER READING

Folkman, J., Watson, K., Ingber, D., et al., 1989. Induction of angiogenesis during the transition from hyperplasia to neoplasia. Nature 339, 58–61.

Gros, S.J., Dohrmann, T., Peldschus, K., et al., 2010. Complementary use of fluorescence and magnetic resonance imaging of metastatic esophageal cancer in a novel orthotopic mouse model. Int. J. Cancer 126, 2671–2681.

Hanahan, D., Folkman, J., 1986. Patterns and emerging mechanisms of the angiogenic switch during tumorigenesis. Cell 86, 353–364.

Nozawa, H., Chiu, C., Hanahan, D., 2006. Infiltrating neutrophils mediate the initial angiogenic switch in a mouse model of multistage carcinogenesis. Proc. Natl. Acad. Sci. U.S.A. 103, 12493–12498.

O'Reilly, M.S., Boehm, T., Shing, Y., et al., 1997. Endostatin: an endogenous inhibitor of angiogenesis and tumor growth. Cell 88, 277–285.

CONCLUDING REMARKS

Jain et al (1997) indicated nine different criteria to define an ideal assay for quantitative angiogenesis: "(1) The release rate and the spatial and temporal concentration distribution of angiogenic factors(s)/inhibitor(s) should be known for generating the dose–response curves. (2) If neoplastic cells are used as a source of angiogenic factors, they should be genetically well defined in terms of oncogene expression and production of growth factors. (3) The assay should provide a quantitative measure of the structure of the new vasculature. (4) It should provide a quantitative measure of the functional characteristics of the new vasculature. (5) There should be a clear distinction between newly formed and preexisting host vessels. (6) Tissue damage should be avoided, since it may lead to the formation of new vessels. (7) Any response seen in vitro should be confirmed in vivo. (8) Such an assay should permit long-term and, if possible, noninvasive monitoring. (9) It should be cost-effective, rapid, easy to use, reproducible, and reliable."

The different mechanisms of angiogenesis are investigated destructively by different models, which can be divided into in vitro and in vivo models. The first ones use endothelial cells, isolated from capillaries or vessels, or vascularized tissue such as rings of the rat aorta. The cells or tissue are embedded in gels such as a Matrigel and the differentiation is evaluated qualitatively or quantitatively. However, since the response in vitro may differ from the one in the organism, in vivo models are developed.

Each experimental approach offers specific advantages and suffers certain limitations. The timing, concentration, location, and mode of treatment administration are critical in all in vivo assays. Novel angiogenesis targeted therapies lack in vivo screening models suitable for objective, quantitative pre-clinical testing, making it difficult to obtain dose–response analysis (Folkman et al., 2001).

In this context, the divergent outcomes obtained from in vivo angiogenesis assays are not unexpected. Therefore, a combination of more than one assay may be necessary in order to determine the pro- or antiangiogenic effect of a substance on endothelial cells.

REFERENCES

Folkman, J., Browder, T., Palmblad, J., 2001. Angiogenesis research: guidelines, for translation to clinical application. Thromb. Hemost. 86, 23–33.

Jain, R.K., Schlenger, K., Hockel, M., et al., 1997. Quantitative angiogenesis assays: progress and problems. Nat. Med. 3, 1203–1208.

Printed in the United States
By Bookmasters